EQUINOX
SPACE

JACK CHALLONER

First published in 2000 by Channel 4 Books, an imprint of Macmillan
Publishers Ltd, 25 Eccleston Place, London SW1W 9NF,
Basingstoke and Oxford

www.macmillan.co.uk

Associated companies throughout the world.

ISBN 0 7522 2331 3

9 7 5 3 1 2 4 6 8

A CIP catalogue record for this book is available from the British Library.

Design by Jane Coney
Typeset by Ferdinand Pageworks
Printed in Great Britain by Mackays of Chatham plc

ACKNOWLEDGEMENTS

Many people have helped me to write this book. Friends and family – in particular Paula James, Carolyn McGoldrick, Daniel Brookman, Karen Darling and Jilly Duckworth – have been supportive as ever. I would like to thank Emma Tait at Channel 4 Books, and my editor Christine King, for their professionalism and enthusiasm. I would also like to acknowledge the makers of the *Equinox* series for communicating issues of contemporary science and contributing to the public understanding of the subject.

PRODUCTION CREDITS

Day Return to Space
accompanies the *Equinox* programme of the same name made by TV6 (Scotland) Limited for Channel 4.
First broadcast: 19 July 1999

What Shall we do with the Moon?
accompanies the *Equinox* programme of the same name made by Wall to Wall Productions for Channel 4.
First broadcast: 19 July 1999

Space – The Final Junkyard
accompanies the *Equinox* programme of the same name made by Eagle & Eagle Limited for Channel 4.
First broadcast: 13 September 1999

Sun Storm
accompanies the *Equinox* programme of the same name made by Eagle & Eagle Limited for Channel 4.
First broadcast: 31 July 1999

Black Holes
accompanies the *Equinox* programme of the same name made by Pioneer Film & TV Productions for Channel 4.
First broadcast: 8 September 1997

The Rubber Universe
accompanies the *Equinox* programme of the same name made by Union Pictures Limited for Channel 4.
First broadcast: 5 June 1996

CONTENTS

INTRODUCTION

Space is where we live, and 'Earth' is our home address. From our beautiful planet, we gaze out into space – in awe of the tantalizing beauty that lurks there and the grand scale of it all. Over the past few decades we have actually begun to venture into space. But the more we look, the more we see there is to understand; and the farther away we explore, the farther we want to go.

Space attracts us for many reasons. Some people are engaged in a quest for understanding, to help make sense of our existence. Some are determined to conquer space, to expand human frontiers. Others see space in terms of its commercial opportunities. All these reasons are investigated in this book, each chapter of which is based on subjects covered by Channel 4's *Equinox* series. Here, I have been able to extend the scope of the television programmes, and to present the science, history, philosophy and politics of these stories in greater detail.

The study of space can be conveniently divided into astronomy and space exploration. Astronomy involves observation, scientific thinking and experimentation, and aims to discover the nature of space and the things it contains. Space exploration involves developing technologies that can escape Earth's gravity and control spacecraft in Earth orbit and beyond. There is often no clear distinction between astronomy and space exploration. Modern astronomy relies on space exploration to enable humans or robot probes to collect information from space. And space exploration requires knowledge about the space environment, as well as knowledge about the nature of space destinations – whether those are the Moon, the planets or, eventually perhaps, the stars.

A major part of astronomy is astrophysics – the study of the physical processes behind the things we see in space. Astrophysics involves knowledge of mathematics, of gravity, of the behaviour of light and other forms of electromagnetic radiation, and an understanding of nuclear reactions, of chemistry and of the nature of space and time. Astronautics is the science and technology of the construction and operation of spacecraft. Much of what is involved in the actual theory and practice of modern astronomy and space exploration is beyond most people's comprehension or interest; but the discoveries of astronomy can have profound effects on everyone. How does it feel to know that we live on an extraordinary but cosmically insignificant planet in orbit around an ordinary star in an ordinary galaxy?

What people see when they gaze into space is a fantastic array of lights – the Sun, the Moon, the planets, stars and galaxies, comets and shooting stars – all shifting around at different speeds, so that the sky is never exactly the same twice. Some of these objects are relatively close to Earth, while others are unbelievably far away. Meteors (shooting stars) are caused by rocks and dust burning up as they enter our atmosphere. The light they produce takes only a tiny fraction of a second to travel the few tens of kilometres from the upper atmosphere to the ground. Light from the Moon takes less than two seconds to reach us. The Sun is about eight 'light minutes' away, while a planet can be between several light minutes and several light hours away, depending on which planet it is and where it is in its orbit. The stars are much farther away – light from the nearest star takes just over four years to reach us. All the stars that people can see with the unaided eye are just a small part of a huge collection: a galaxy, the Milky Way. Ours is a typical galaxy: like an island of stars, 100,000 light years across, in the vast empty ocean of space. The most distant object visible to the unaided eye – as a small, faint, fuzzy patch of light on a dark, clear night – is another of these island communities, called the Andromeda Galaxy. Light from there takes more than two million years to reach Earth. The very farthest objects so far detected in space lie an unbelievable 10,000 million light years away.

Our efforts in the exploration of space have resulted in some spectacular achievements. From Earth's orbit, astronauts have seen our planet from an unprecedented viewpoint, and uncrewed Earth-orbiting observatories such as the Hubble Space Telescope are much better placed for looking into space than their ground-based counterparts. Space exploration is not designed only to give astronauts an exciting ride or to help scientists to look deeper into space: it can bring benefits to many people on Earth, beyond the scientific community. Communications satellites provide international television, telephone and Internet links; there are satellites to help predict the weather, and satellites to help people to navigate; there are even military satellites, secretly doing whatever they do in case the other side does it first. But Earth orbit can be just the departure lounge for travel farther afield. Spacecraft can remain in orbit without using any fuel, and it is relatively easy to blast off from Earth orbit to the Moon, the other planets, and comets and asteroids. Human space pioneers have made it as far as the

Moon, but uncrewed robot probes have visited all the planets except for Pluto, the farthest from the Sun. And even the uncrewed space probes have travelled only tiny distances compared to the size of our galaxy. In the foreseeable future, the number of journeys we undertake is likely to increase, rather than the distance of these journeys. Space travel will become more widespread, more routine.

The earliest people probably had no dreams of travelling into space, because they did not understand what and where space was. They were probably too busy surviving to worry about discerning the nature of the Universe, although it is nice to think they at least looked up in wonder. The equivalent of astrophysics for the ancients was workable myths that involved the gods who were in charge of the heavens and Earth. When stable civilizations formed, they began to observe how the Sun, planets and stars shifted across the sky, and learned to construct calendars that would help them to know when to plant their crops. These people used the Sun, Moon and stars as aids to navigation, too. In ancient India and Egypt, astronomers used complicated mathematics to predict when solar and lunar eclipses would happen. The ancient Greeks went further, forming theories that attempted to explain what the Sun, Moon, planets and stars actually are. The prevailing view was that the Earth was at the centre of the Universe, with the Sun, Moon, planets and stars exist on a celestial sphere that rotated around it. Complicated variations of this concept were put forward, which involved spheres within spheres in an attempt to explain the way the planets were seen to move across the sky.

The idea of an Earth-centred Universe was convincing enough until the time of the Renaissance, when the long-standing ideas of the ancient Greeks came under serious scrutiny and re-evaluation. In the sixteenth century, the Polish astronomer Nicolaus Copernicus proposed that the Sun, and not the Earth, was at the centre of the Universe. To many people – particularly leaders of the church, who could not accept that the Earth was not the centre of everything – this idea was not easy to swallow. But several important new discoveries lent support to it. The German astronomer Johannes Kepler worked out mathematically how planets move in their orbits, and his findings corresponded to a Universe like the one Copernicus had envisioned. The Danish astronomer Tycho Brahe discovered a new star in the sky (a supernova), which challenged the idea that the heavens were perfect and unchanging celestial spheres. The Italian scientist Galileo discovered four objects that circle Jupiter in the same way as Copernicus had suggested the planets orbit the Sun. Galileo had also seen surface details on the planets – they were not perfect and unchanging. And perhaps the most convincing evidence of all in favour of Copernicus's view of the Universe came from Isaac Newton's theory of gravitation. Newton realized that gravity could be the mechanism by which Copernicus's Universe could work. Gravity between the Sun, the Moon and the planets explained faithfully their observed movements.

Newton's theory was the beginning of astrophysics proper, because it provided a framework upon which to investigate what the Sun and the planets are: how big they are, how far away and how fast they are moving, for example. And it secured the demise of the Earth-centred views of the Universe.

The telescope led to discoveries of previously unknown planets, and all manner of beautiful fuzzy objects – some of which are now known to be clouds of gas (nebulas) in our own galaxy, while others are now known to be galaxies separate from our own. Using telescopes, astronomers discovered moons around Saturn, Mars and Uranus. The nature of our space neighbourhood, the Solar System, was being uncovered, but the stars and the fuzzy objects remained a mystery. In 1838 the Universe got bigger: the German astronomer Friedrich Bessel made the first measurements of the enormous distances to the stars. Such measurements enabled astronomers to work out the distribution of stars in what they thought was the Universe, though the nature of the fuzzy objects remained a mystery. In 1924 the Universe got bigger again: American astronomer Edwin Hubble worked out the distance to one of the fuzzy objects – the Andromeda Galaxy – and found that it was much farther away than any of the stars. Hubble realized that what astronomers had believed to be the entire Universe was just one of a large number of galaxies.

Many scientific disciplines have contributed to the development of astronomers' ideas. For example, spectroscopy – the analysis of the spectrum of light from stars – revealed what stars are made of, and nuclear physics explained how stars produce their massive energy output. Also, the discovery of infrared, ultraviolet and radio waves opened new windows on the Universe, extending the scope of astronomical observations and leading to the discovery of new types of astronomical object. Photography has been another vital tool for the astronomer: long-exposure photographs – or today, modern electronic detectors – reveal new objects too faint to see otherwise with or without a telescope.

Compared to astronomy, the exploration of space has a very short history. Novels and scientific papers on spaceflight began to appear in earnest around 1900, and rapid development of rockets fit for spaceflight took place from the 1920s. Space was first reached at the end of the 1940s, but a spacecraft first went into orbit around the Earth in 1957. This was the beginning of the space age. As well as launching telecommunications satellites and achieving crewed spaceflight, space scientists and engineers sent probes to the Moon, Mercury, Venus and Mars, and later to Jupiter, Saturn, Uranus and Neptune. These probes – together with orbiting space telescopes – have brought the planets into our homes, presenting for the first time really close-up photographs of them. As far as launching satellites and space probes is concerned, spacegoing technology has become routine.

Initial developments in space technology were rapid. At the time of the first Apollo Moon landings, people had great hopes that space travel would be affordable and routine within their lifetimes. But the early developments were driven more by Cold War politics than by the quest for knowledge about the Universe or the drive towards space exploration. The Americans and the Russians were engaged in a space race. In 'Day Return to Space', we find out what attempts space engineers are making to develop cheaper, reusable spacecraft. Many private companies are becoming impatient with the attempts being made by government space agencies in this direction, and are taking it upon themselves to exploit space commercially. Space is already becoming the next great business opportunity, and tourism is sure to be a part of it – in 'What Shall we do with the Moon?' we discover that hotels are among the money-making opportunities planned for the Moon. There is a cautionary tale in 'Space – the Final Junkyard', which tells of the consequences of irresponsible use of outer space. Orbiting our planet at dangerous speeds are vast clouds of debris from previous space missions, including malfunctioning nuclear-powered satellites. There are many more satellite launches planned – to coincide with and help support the dramatic increase in global telecommunications. Space is becoming congested and dangerous, and we are becoming more dependent on space in our daily lives with every year that passes.

Some of the things upon which we have come to depend may be under threat not only from fast-moving space debris, but also from powerful magnetic storms on the Sun. These sun storms can cause power cuts and perhaps endanger health on Earth, too, and are the subject of 'Sun Storms'. There is a massive worldwide effort to develop our understanding of the Sun, so that we might be able to forecast the 'space weather'.

Astronomy is normally about observing things in the sky – by looking at electromagnetic radiation, such as light, coming from them. But astronomers hunting the most elusive of celestial objects – black holes – have to use some cunning methods in their search. They have recently shown that these mind-boggling objects, once part of science fiction and abstract theory, almost certainly exist. 'Black Holes' tells their story. It may be that the space and time of our Universe was created by the explosion of something like a black hole. That explosion – the Big Bang – happened a long time ago, but just how long ago is a hotly debated topic, and is explored in 'The Rubber Universe'.

Finally, a brief Afterword looks ahead to the fascinating future developments in the study, the exploration and perhaps the exploitation of space.

DAY RETURN TO SPACE
... searching for the space Volkswagen...

The space race of the 1950s and 1960s inspired a generation with the idea that within its lifetime there would be space hotels and bases on the Moon and Mars. Initially, progress in space technology was extremely rapid: the first artificial satellite was launched into orbit in 1957; the first humans made it into space in 1961; and people were standing on the Moon by 1969. Since then, most journeys into space have been undertaken by uncrewed space probes or Earth-orbiting satellites. The space hotels and Moon bases have remained a distant dream, still firmly rooted in science fiction. Will travel into space ever be as routine as long-haul air travel is today? Will *you* ever get the chance to go into space?

Orbital adventures
If a holiday in space were as affordable and available as a cruise on the *QE2* or a month on safari in Africa, would you consider going? Would you like to experience the weightlessness that being in orbit brings? Would you want to see our beautiful blue and white planet beneath you, and a glorious sunrise every hour and a half? Would you want to gaze at the stars from an observation post in a space hotel high above the atmosphere? And if the price was right, would you one day want to go and spend some time on the Moon?

Many people say 'yes' to this kind of question: surveys have shown that between 60 and 70 per cent of people in the USA and Japan would like to take a trip into space. One day, space travel will be a major part of the tourism industry, and will therefore have to be as routine as long-haul aeroplane flights. In the past few decades, many feasibility studies have been carried out into the possibility of space tourism. Most of the recent studies suggest that it would be easy to realize space tourism as a going concern, and they estimate that tourists will have to pay between $10,000 and $100,000 per trip. Despite the perceived commercial value of space tourism, government-run space agencies such as NASA (National Aeronautics and Space Administration) have so far done little to promote the idea. According to a 1998 NASA report, *General Public Space Travel and Tourism*: 'US private sector business revenues in the space information area now approximate $10 billion per year, and are increasing rapidly. Not so in the human spaceflight area. After spending $100s of billions in public funds thereon, and continuing to spend over $5 billion per year, the government is still the only customer for human spaceflight goods and services.'

Government-led space programmes are not making any headway in the space tourism business – which might go the same way as the Internet. In the late 1960s and through the 1970s the Internet was limited to the international scientific community and the US defence agencies. But since the early 1980s access to this growing global information exchange has been possible for anyone with a computer and a modem. Progress in the technology has been astonishingly fast, and now hundreds of millions of people use it to find information, keep in touch quickly and easily, watch videos and listen to the radio. What made this possible, in such a short space of time, was private enterprise. As soon as it became possible for people to make money from this world-wide computer network, people found the best ways to do just that. Technology has made the Internet available, in principle, to everyone, but the development of that technology has been driven by the desire to make money. With private finance, space travel could develop in the same way as the Internet did, quickly becoming accessible to more and more people.

Private organizations – with names such as Spacetopia Incorporated, Zegrahm Space Voyages and Virgin Galactic Airways – have already been formed with space tourism in mind. In March 1997 a German company, Space Tours Gmbh, hosted the first International Symposium on Space Tourism. In attendance were representatives of the aerospace industry, hoteliers and financiers. Several companies presented well-developed plans for tourism in space. That same year, the Japanese construction company Shimusu announced its plans to open a space hotel by the year 2020 – offering anyone with enough money the chance to take a holiday at an altitude of 450 kilometres. And in 1998 a US company called Space Islands announced another well-developed and ingenious plan, for taking holidaymakers up to a hotel 320 kilometres above the Earth's surface. This hotel would be built from discarded Space Shuttle fuel tanks, which at present are jettisoned and burn up as they re-enter Earth's atmosphere. Each of the tanks is 47 metres long, and the proposal is to join twelve of them together to make a huge wheel with a circumference of nearly 600 metres and a diameter of 179 metres. The tanks, on the rim of the wheel, would contain luxury accommodation for up to 350 people. The wheel would slowly rotate, simulating gravitational forces inside the tanks equivalent to one-third that experienced on Earth's surface. For the hub of the wheel there would be a module where people could experience zero gravity. The company hopes to offer tourists a chance to visit their hotel as early as 2005.

This idea may not be pie in the sky: Hilton Hotels, Virgin Airways and British Airways have all expressed an interest in becoming involved. Space Islands estimates that by the year 2010, the cost of a week's holiday at the hotel may be as low as $15,000. With companies selling advance tickets, space tourism is already big business. But there is one thing that is holding it back: a cost-effective vehicle that can launch people into space.

At present, the pinnacle of person-carrying spacecraft technology is NASA's Space Shuttle (more formally known as the Space Transportation System – STS). This remarkable vehicle was tested in the atmosphere in 1981, and made its first orbital flight – in space – in 1983. The Space Shuttle was designed to be reusable and relatively cheap, to make launching into space a routine procedure. And yet each launch costs up to $500 million. NASA's chief administrator, Daniel Goldin, explains the problem: 'We, NASA, are limited in what we can do in space because we spend too much of our budget on launch. Of a budget of $13.8 billion, we're spending close to $5 billion on launch. Think what we could do with that money.' Several thousand members of ground crew are needed for each launch of the Space Shuttle, which can carry up to only seven highly trained astronauts.

To understand why rocket launches are so expensive, it helps to look back over the history of rocket flight.

Countdown to the space age
The first object propelled into space was an American rocket, the V-2-WAC-Corporal combination rocket, which rose to a

height of 390 kilometres in 1949. It consisted of a V-2 rocket – of the same design as those launched against several European countries during World War II – with an American wartime rocket attached to the top. The highest altitude reached before 1949 was 85 kilometres, by a German V-2 rocket in 1942. According to space scientists, this is 15 kilometres short of space: they generally define the boundary between Earth and space as 100 kilometres above the Earth's surface. There is no distinct boundary, however – just a gradual decrease in the density of the atmosphere with increasing altitude. At 100 kilometres, the air is so rarefied that no jet engines can function. Jet engines take in air, which contains oxygen that their fuel needs if it is to burn. Rockets, on the other hand, must work in extremely rarefied air or in a vacuum, so they carry both their fuel and oxygen with them.

A rocket accelerates a spacecraft because it produces a huge thrust. The source of this thrust is the expulsion of gases at high speed from the back of the rocket. An inflated party balloon can produce thrust in this way, as air escapes through its neck. As it pushes air backwards, the balloon accelerates forwards. This general principle – that pushing something in one direction will push you in the opposite direction – is a consequence of the laws of motion discovered by Isaac Newton. For a graphic illustration of Newton's Laws, imagine that you are on an ice-skating rink, wearing ice skates, and that next to you is a heavy metal safe standing firm with its legs stuck into the ice. If you push hard on the safe, it stays still while you move off in the opposite direction. The fact that you start moving means that a force is being exerted on you, and it is the safe that is exerting this force. This is an example of a reaction force – an equal and opposite counterpart to a force exerted on an object. The reason you can sit on a chair without falling through it is that the chair pushes upwards on you with a force equal and opposite to your weight, which is pulling you downwards towards the ground. If the chair suddenly disappears, then so does this balancing reaction force, and you fall down to the next lowest object that will provide a reaction force: the ground.

It is not just objects pushing against each other that produce equal and opposite forces in this way. Any pulling force also has its equal and opposite counterpart. When you pull a sled over snow, the sled pulls back on you – by virtue of tension in the rope. As there is less friction between its runners and the snow than between your feet and the snow, the sled moves. If the sled was stuck firm in the snow, the tension in the rope would make you move, however. Objects do not even have to be in contact to exert equal and opposite forces on each other. A paper clip attracts a magnet with as much force as a magnet attracts a paper clip, for example. In the same way, you exert the same force upwards on the Earth as it exerts on you.

In the ice-rink scenario, imagine now that the safe is turned on its side, so that there is very little friction between it and the ice. When you push on it this time, the safe moves away from you. But there is still a reaction force – the safe pushes on you as before – so you start moving, too, in the opposite direction. The mass of the safe is important: a small plastic safe will speed away from you when you push it on the ice rink, while you hardly move. Conversely, a 20-tonne safe would practically stay still, but in this case it would be you moving away at high speed. In fact, if you multiply your mass by the speed you achieve by pushing against the safe, you will get exactly the same answer as if you multiply the mass of the safe by the speed it moves away in the opposite direction.

So much for ice rinks and safes. How does this relate to the party balloon? It seems as though there is nothing pushing the balloon, and yet the balloon shoots forward faster and faster, until all the air has been expelled. The air is expelled because it is pushed by the stretched rubber as the rubber shrinks to regain its normal size. And so, according to Newton's laws, the air exerts a force on the balloon, too. The balloon accelerates until the rubber is no longer stretched. This does not quite explain how a firework rocket works, since there is no stretched rubber to force the air out. In this case, gunpowder explodes inside a rigid cardboard casing. Gunpowder is a solid mixture that includes an oxygen-containing substance. Oxygen is necessary for the other components of the mixture to burn. Because oxygen is part of the gunpowder mixture, and is released as the gunpowder heats up, the gunpowder does not have to take oxygen from the surrounding air and the mixture burns very quickly. As it does so, it produces hot gases. These gases take up more volume than the solid gunpowder, and force their way out at the back of the paper cylinder. The rocket shoots up into the sky.

Rockets have been in existence for about a thousand years, since they were used in firework displays in ancient China. In the thirteenth century, Chinese armies used what they called 'fire arrows' against the invading Mongols. Rocket technology in Europe had a slow development: having arrived in the thirteenth century, rockets were not used effectively in battle until the end of the seventeenth century. All early rockets contained gunpowder in a pointed paper tube. During the eighteenth century, the efficiency of rockets was improved by housing the fuel in a metal tube. In the same century, these rockets were used in warfare in India, against the British Army. The first reliable system designed for military use was developed by a British weapons expert, William Congreve. This was used in 1812, during the Napoleonic wars, when Congreve's rockets rained on several European cities. Congreve's design had an important feature: a long stick that stabilized the rocket in flight. This stability was improved further by a British engineer, William Hale, who designed rockets with angled exhaust pipes that made the rockets spin as they pushed their way along. Many modern rockets still use this kind of spin stability.

The first person to suggest seriously that rockets could be used to launch people into space was a Russian school teacher, Konstantin Tsiolkovsky. In his 1895 book *Dreams of Earth and Sky*, he set out his visionary ideas: for example, he foresaw the possibility of satellites, the use of solar power to provide energy during space travel, and the eventual spread of human civilization across the entire Solar System. In 1898 he worked out many of the principles behind the use of rockets for travel into space and, in 1903, he published a book called *Exploration of Space by Reaction Apparatuses*, in which he wrote: 'Earth is the cradle of the mind, but one cannot live in the cradle for ever.' Tsiolkovsky developed his ideas in many other articles, hoping to inspire others to share his dream. In Russia, he became famous, and was given funds to continue his research into astronautics and aeronautics. But he was not well known outside his own country until long after he died, in 1935.

The first liquid fuel rocket was built by American rocket pioneer Robert Goddard in 1926. Until then, all rockets had solid fuel, often compacted powders as in a firework. Goddard's first liquid fuel rocket climbed to an altitude of 13 metres, in a flight that lasted just two and a half seconds. Liquid fuel rockets have several advantages over solid fuel rockets, particularly for space travel. One of the main advantages is controllability. In a solid rocket, the fuel burns at a set rate. This rate can vary through the rocket's ascent, but only by pre-packing the fuel in a particular arrangement. In a liquid fuel rocket, pumps are used to deliver the fuel to the engine, and therefore control at what rate it burns. By regulating the pumps, you can control the rocket's thrust, even after launch. Like Tsiolkovsky, Goddard was a genuine visionary. As a school physics teacher in Worcester, Massachusetts, he included potential visits to the Moon in the curriculum. He carried out his own research, including proving for the first time that a rocket would work in the vacuum of space – many people believed that a rocket would not work in space because there is no air against which the rocket can push. Initial funding for his work came from the Smithsonian Institution, which in 1920 made public the details of his funding application, and Goddard became mockingly known as 'The Moon Rocket Man'. In 1929 one newspaper featured the headline 'Moon Rocket misses target by 238,799 miles'. Goddard continued his work, undeterred. In 1935 one of his liquid fuel rockets became the first manufactured, self-propelled object to travel faster than the speed of sound.

Also during the 1930s another rocket pioneer – this time in Germany – was pushing back the frontiers of rocket design for the Nazi war effort. He was Werner von Braun, designer of the V-2 rockets that brought terror to Britain, France and Belgium during 1944 and 1945. After the end of the war, von Braun and about a hundred of his colleagues moved to the USA so that they could continue their work on rocketry. The rest of the German rocketry researchers and engineers moved to Russia. Both superpowers had existing rocket programmes, designed to launch intercontinental ballistic missiles with nuclear warheads. They used the expertise and experience of the members of the German team to make their rockets fly farther and higher than ever before. The rocket that reached 390 kilometres in 1949 was a combination of the V-2 and an existing US Army rocket. Seven years after the V-2-WAC-Corporal crossed the boundary into space, another American rocket reached an even higher altitude of 1,090 kilometres.

However, all these objects fell back to Earth again, just like stones thrown into the air. If you could throw a stone with a speed greater than eleven kilometres per second, it would never fall down again. Instead it would continue travelling away for ever – this is not too desirable for a satellite or crewed space vehicle. What you want to do with satellites is to stop them from falling to Earth, but also to keep them from travelling away for ever. If you simply place a spacecraft in space, stationary, it will fall to Earth: it will be attracted back down to the ground again by gravity. The gravitational force on an object decreases gradually with altitude; despite popular belief, there is gravity in space. Even if the satellite was at an altitude of several thousand kilometres, it would – eventually – fall back to Earth. To get around this problem – to prevent falling back to Earth at all – the satellite must orbit the Earth. In this case, it will still be pulled towards the Earth, but it will never move any closer to it.

The gravitational pull between the Earth and an orbiting satellite is just enough to keep the satellite in orbit. If the satellite moves too slowly, it will be pulled back to Earth. If it moves too fast, it will shoot off into space. Out in the depths of space, it will either be pulled back to Earth again or continue moving away for ever. And if it is pulled back to Earth, the satellite will shoot past again only to be pulled back again. It will be moving in a very unusual – or 'eccentric' – orbit. To move in a practical orbit, where its distance from Earth does not vary too much, the satellite has to travel at just the right speed. This speed depends upon the altitude. At an altitude of 300 kilometres, it must be travelling at nearly eight kilometres per second to stay in orbit. It takes about an hour and a half to complete each orbit at this altitude. The orbital speed decreases as it moves farther away from the Earth. At an altitude of 50,000 kilometres it has to travel at only 2.6 kilometres per second. This is still more than four times the top speed of Concorde, and the satellite would be a long way from home. At this distance, it would take about thirty-eight hours to complete each orbit. Most communications satellites orbit above the equator at an altitude of 35,900 kilometres. At this distance, each orbit takes twenty-four hours, and the satellite remains above the same point on Earth – it is geostationary.

Staging a comeback

Any object in orbit around another is a satellite. The Moon is our only natural satellite, but there are now many thousands of artificial ones. Crewed spacecraft generally orbit at altitudes of less than 480 kilometres, in a 'low Earth orbit'. The people-carrying part of the Space Shuttle – called the orbiter – flies to low Earth orbit, stays there for between ten and sixteen days, and then returns to Earth. It was designed to be the first reusable spacecraft. However, not all of it is reusable: to reach the speed necessary to move into orbit, solid rocket boosters are needed in addition to the spacecraft's main engines. At an altitude of 45 kilometres, the solid rocket boosters have used all their fuel and they are jettisoned, recovered and used again up to twenty times. The Shuttle's main engines rely on fuel from a large external tank, which remains attached until orbit, when it too is jettisoned. The empty fuel tank is left to re-enter the atmosphere and burn up (although it could always be salvaged and used to build a space hotel).

So, like all missions that have so far made it to orbit, the Space Shuttle is a multi-stage launch vehicle, and only a small proportion of the mass of the spacecraft at launch makes it back again. This is wasteful and is always going to be costly, as Daniel Goldin of NASA explains: 'Do you think we could build an aircraft to go from New York to London and have the aircraft take off and then a piece drops off and then another piece? That's why it's inefficient.' The cost of carrying people into space will remain prohibitively high until launches become cheaper. And what is needed to make it cheaper is a totally reusable space vehicle: a SSTO (single-stage-to-orbit) spacecraft.

The multi-stage approach to launching spacecraft is a remnant of the early days of space travel, when it did not matter how cheaply you could get into space – just that you got there at all. The launch vehicles looked and worked like missiles, because the American and Russian space programmes both had their origins in the development of warhead-carrying rockets. As well as the development of missiles, several scientific missions were considered. As part of the International Geophysical Year – which strangely ran from July 1957 until December 1958 – the Americans and the Russians both planned to send up satellites that could carry out Earth science experiments from the ultimate vantage point, in orbit. American plans hinged on the Vanguard rocket, which had been put together by a team from the US Navy. During 1956 and 1957, however, attempts to launch the Vanguard ended in disaster. Still, the Americans smugly assumed that they were far ahead of the

Russians in spaceflight technology. But they were wrong: the first manufactured object to go into orbit was Russian.

After several announcements of their intent – which most Americans wrote off as unrealistic – Russian space scientists succeeded in placing an artificial satellite into orbit around the Earth on 4 October 1957. This was Sputnik I – a silver-coloured metal sphere with four long straight aerials and a mass of 83.5 kilograms. The launch of Sputnik I was originally planned to coincide with the hundredth anniversary of the birth of Konstantin Tsiolkovsky, which would have been 19 September 1957; the delay was due to technical problems. Sputnik's aerials transmitted a test signal for ninety-two days, after which the satellite burned up as it re-entered the atmosphere.

Scanning the sky with radio telescopes, to tune into the test signal, was the only way to convince some people that Sputnik was really in orbit. Once proven, the Russian achievement dealt a blow to the national pride of the Americans. Not only was it a matter of pride: the Americans and Russians had been involved in diplomatic rivalry – the Cold War – since shortly after the end of World War II. If Russian rockets could manage to place a satellite into orbit, they could certainly launch a nuclear warhead to strike anywhere on Earth. As Daniel Goldin remarks: 'The objective of the space programme in the Eastern Bloc and the Western Bloc was to show technical superiority … it was about winning the hearts and minds of the uncommitted countries of this world.' And the Americans were lagging behind.

Things got worse before they got better for the American space programme: in November 1957 Sputnik II, weighing more than half a tonne – nearly six times as heavy as Sputnik I – was placed successfully into orbit. Sputnik II carried a dog, called Laika, around the Earth for 162 days, until the dog and the satellite both burned up as they re-entered the atmosphere. At the time, the American satellite programme was in tatters. The Navy's Vanguard project included a plan to send a small scientific payload into orbit, but the Vanguard rockets were not working as hoped. Fortunately for the USA, Werner von Braun had been developing another launch system, the Jupiter-C, with the US Army. And on 31 January 1958 a Jupiter-C rocket carried a smaller rocket and a payload into space, where the small rocket pushed the payload – a satellite called Explorer I – into orbit. Explorer I was considerably smaller than even the first Sputnik, but did help to make important scientific discoveries, such as the existence of a belt of charged particles encircling the Earth, called the Van Allen Belts. Further American satellites followed, including one launched on top of – at last – a Vanguard rocket in March 1958.

The space race was now in full swing. The next major challenge was to send space probes beyond Earth's orbit to investigate the Moon and the other planets. Again, the Americans lagged behind. They attempted to send three probes to the Moon late in 1958, but the rockets that were to take them into orbit failed on launch. And so the Russian space effort made it to the Moon first, too. In 1959 they sent up two probes: Luna II deliberately crash-landed on the Moon's surface, while Luna III took photographs of the far side of the Moon. As the Moon turns once on its axis in exactly the same time it takes it to orbit, one particular half of its surface is always hidden from Earth, so the Russian achievement was significant. The American space programme did pick up, however, particularly as a result of the formation of NASA in October 1958, in response to the launch of Sputnik I. Until 1958, American efforts to travel into space resided in NACA (National Advisory Committee for Aeronautics) and the disparate plans of the US Army and Navy. NASA provided a way of administering and co-ordinating the American efforts in space.

The next obvious development, after satellites and space probes, was to fire a person into space. Sputnik III had carried a dog, and in 1960 another Russian craft, Sputnik V, carried two dogs – Strelka and Belka – into orbit and safely back again. This was obviously an important break-through: Sputnik V was the first object to be recovered from orbit, and the fact that the dogs had returned alive made this a vital step towards putting humans into space. The first person ever to make it into space was Russian astro-naut Yuri Gagarin, who made slightly more than one orbit of the Earth in April 1961 at a maximum altitude of 300 kilometres, before re-entering the atmosphere to a hero's welcome.

The Americans soon followed, in May of the same year, sending Alan Sheppard into space – though not into orbit. Also in this month the American space programme received an important boost, when President John F. Kennedy announced an effort to redress the balance in the space race: he committed the US space programme to putting people on the Moon, with the Apollo programme. When elected in 1960, Kennedy had made crewed spaceflight a key political goal; now, on 25 May 1961, he proclaimed: 'I believe this nation should commit itself to achieving the goal, before this decade is out, of landing a man on the Moon and returning him safely to Earth. No single space project in this period will be more impressive to mankind, or more important in the long-range exploration of space; and none will be so difficult or expensive to accomplish.'

In February 1962 an American rocket took John Glenn into orbit, as part of the project Mercury, the precursor to the Apollo programme. To achieve the goal of Apollo within the proposed time scale, more than four per cent of the entire US federal budget would have to be committed to the project. The Saturn V rockets that launched the Apollo spacecraft into space were incredibly powerful, but they were also the ultimate in wastefulness. Single-stage launch vehicles would have been much cheaper, particularly if they could come down in one piece, to be reused. But cost was not important, since space programmes in the early years were driven largely by Russian and American political ambitions, with little care paid to the costs involved. If cost had been an issue, development in the space programmes would not have been as rapid. And so it was huge, expensive, missile-shaped rockets that made headway in the space race.

The most obvious way to make a space vehicle that is reusable is to design it like an aeroplane, so that it can land horizontally, on an airfield. The Space Shuttle is a compromise between a reusable spaceplane and a wasteful multi-stage rocket. So, although it is lifted into space by huge rockets and an external fuel tank that are jettisoned during the mission, it does land back on Earth gracefully, just like a glider, to be reused. In 1983, the year of the Space Shuttle's first mission, British aerospace engineer Alan Bond proposed a radical spaceplane, called HOTOL (horizontal take-off and landing). This would have been an elegant spacecraft, with swept-back wings and an engine that would work as a jet when it was in the atmosphere, but as a rocket when in space. Much less oxygen would have to be carried in the fuel tanks, since the engine would take oxygen from the air for much of its ascent – a significant advantage over conventional rockets. However, to achieve orbital velocity from a standing start on a runway, the craft would have to accelerate quickly to hypersonic speeds. The project was developed by British Aerospace and Rolls-Royce, but the technological and financial challenges were too great and the project was shelved in 1989. However, a private company formed by ex-Rolls-Royce engineers is now working on a successor to the HOTOL concept, called SKY-LON. Like HOTOL, it is a spaceplane that will have a hybrid engine – one that will breathe air in the atmosphere but will carry oxygen for flight in space.

And so the quest for a truly reusable spacecraft continues. Until the Space Shuttle, all spacecraft that returned from space crash-landed – except one. In 1963 a rocket-powered aeroplane set an altitude record of 101 kilometres, just crossing the boundary into space. This was the X-15, the culmination of a secret US project to develop supersonic aeroplanes. The 'X' of the X-programme stands for 'experimental'. The earliest product of this top-secret technological drive for innovation – the X-1 – became the first aeroplane to travel faster than the speed of sound, in 1947. The X-plane programme was run first by NACA, and then by NASA. It was bold and dangerous, and many pilots lost their lives, but much was learned about the effects of travelling at extremely high speeds. It was not the main aim of the programme to design spaceplanes, but the development of the Space Shuttle programme can be traced back to these heady days. Much was learned, for example, from the X-15's successful glide back to Earth from above the boundary of space. Since Yuri Gagarin's first spaceflight, fewer than 400 people have ventured into space. To increase that figure dramatically, the cost of getting into space will have to be reduced considerably. What is needed is the equivalent of the Volkswagen: a people's vehicle for space travel. What will this spacecraft be like?

Any ideas?

In 1995, in an effort to develop true reusability, NASA invited aerospace companies to bid for a project to build a successor to the Space Shuttle. This was a continuation of the X-programme, though it was far from top secret, and the new vehicle would be called the X-33. The new vehicle would have to carry into space the same weight as the Space Shuttle can – 23 tonnes – but be totally reusable. Furthermore, it would have to be developed within a budget of $1,000 million – only twice the cost of a single Space Shuttle launch. Of the three serious contenders for the X-33 contract, one was the company that built the orbiter of the Space Shuttle – Rockwell International. Rockwell's design was a winged spaceplane, not unlike the Space Shuttle itself. To comply with the requirements of the X-33, Rockwell had developed the idea of the Space Shuttle into a spaceplane that needed no boosters or external fuel tank. The operating costs would be less, too, as the number of people on the ground would be reduced to fewer than a hundred.

Another design – and the one hotly tipped as the favourite to win the contract – was submitted by McDonnell Douglas, a company that also had long-standing links with the aerospace industry. McDonnell Douglas's radical design was based on a spacecraft that they were already testing, in conjunction with the US Air Force, called the Delta Clipper. The DC-X – a one-third-scale prototype of the Delta Clipper – had already undergone successful trials in 1993, in the Army's White Sands Missile Range in New Mexico. There were two radical features to the Delta Clipper that helped make it an attractive proposition as a reusable launch vehicle. First, it landed vertically, using rockets to slow its descent in a delicate hovering motion. This soft landing could be achieved at space ports almost anywhere, without needing a long runway as the Space Shuttle does. Second, fewer than twenty people were needed for each launch – a significant reduction compared with the several thousand people needed for each Space Shuttle launch. Major Shell, McDonnell Douglas's spaceplanes test manager, says, 'With the DC-X, the goal was three people to fly it, in the operations centre, with fewer than fifteen people on what we call touch labour.' In addition, the Delta Clipper would be able to fly again within a week of landing.

The DC-X's first test flight was an incredible success. This eye-catching vehicle – which looked like a 13-metre-tall, upside-down ice-cream cone – moved up into the air, hovered for a few minutes, carried out a horizontal 'translational' movement, by tilting slightly, then eased itself slowly back down to the ground. A few metres above the ground, three sturdy legs emerged from the flat base of the craft, and these supported it as it touched down. This unusual design brought a great deal of interest, both within and beyond the space science community. Although it was designed as a military 'space truck', under contract by the Strategic Defense Initiative Organization, the Delta Clipper's designers had wider goals in mind. The craft would ultimately be able to carry military equipment, commercial satellites or tourists into space. In 1995 NASA took the DC-X under its wing, renaming it the DC-XA – the Delta Clipper Experimental Advanced. This is why most people assumed that the DC-X would be the vehicle chosen by NASA for the X-33 contract. And McDonnell Douglas themselves were confident. Their X-33 programme manager, Paul Klevatt, explains the capabilities of the team that was developing the proposal – the same team that designed the DC-X: 'McDonnell Douglas was able to fly the DC-X in twenty-four months and three days, from a piece of paper that was just a blank proposal … The learning is there, the processes are in place.'

The third design put forward for the X-33 programme came from the company that assembled the Hubble Space Telescope: Lockheed Martin. Their design for the X-33 – which they dubbed 'VentureStar' – was unusual. It was a single-stage spaceplane, but it had no wings. The shape of the fuselage was called a lifting body: the lift force that would make it fly was created by airflow over the entire body, not over wings as in a conventional aeroplane. The Space Shuttle is a lifting body: although it does have wings, most of its lift force is gained by virtue of the shape of its fuselage. In the VentureStar, there were no wings at all – just small fins for control and stability. The idea for the VentureStar was the brainchild of one person, Dave Urie, who says: 'A lifting body is a thick delta wing; and so we haven't got rid of wings, we've just changed the shape of the wing so that it makes a better hydrogen container.'

Lifting bodies had been around since the 1950s: six of the secret X-planes were wingless lifting bodies, and they were involved in more than 200 test flights between them. These unlikely-looking aeroplanes were carried underneath bomber aircraft, such as the B-52, and released at high altitude. Lockheed Martin's design would have to take off under its own power. And as it was to be a single-stage vehicle, it would have to take all its fuel with it. In fact, the weight of the fuel would have to be around 90 per cent of the total weight at lift-off. In other words, the weight of the spacecraft and its payload must be only 10 per cent of the total. To address this, it was proposed to make the body out of strong but incredibly light new materials and to employ a new type of engine, called an aerospike. Lockheed

Martin's design – with its white body, black nose cone and black and white fins – looked a little like a penguin. It was a plump, graceful, black and white triangle.

In July 1996 NASA's decision on which of the three designs would be taken up as the X-33 was announced: addressing a live television audience, Al Gore, who was then the Vice President, lifted a cover to reveal a model of the winning design. 'This is the craft that can carry America's dreams aloft and launch our nation into a sparkling new century,' he said enthusiastically. To many people's surprise, it was a model of Lockheed Martin's design that was sitting there; Dave Urie's unusual, but slick, penguin-shaped lifting body had won. The team from McDonnell Douglas, whose DC-X design had been tipped to win, were very surprised, and very disappointed. Soon, after four more weeks of successful flight trials, their disappointment turned to disaster. In the DC-X's testing ground in White Sands, the test vehicle toppled as one of its legs failed on landing. As it hit the ground, the whole spacecraft exploded in a huge and devastating orange fireball. This was to spell the end of McDonnell Douglas's DC-X programme.

Dr Hans Mark, an ex-deputy administrator of NASA, believes that the DC-X would never have met the challenges of the X-33 programme: 'The DC-X was a publicity enterprise really, to heighten the tension. Also the DC-X itself had no technology that had anything to do with solving the problem. The engines were thirty-year-old RL10 engines and the aeroshell was made out of a plastic with a low melting point ... I never had much of a use for it.' Those involved in the DC-X project, however, were very upset about the termination of their development programme because of a single failed leg. Major Shell of McDonnell Douglas's Philips Laboratory was one of those for whom this was a bitter blow: 'We put so much of our time, effort and careers into it, we hate to see it end like that.' Daniel Goldin of NASA responded to criticism of the decision: 'They lost; we ran a fair and open competition. This is how democracy works.'

NASA chose Lockheed Martin's design to be the X-33 because it was the one with the most technological challenges to be overcome – the project was the one most likely to drive aeronautics forward. Within weeks of the X-33 announcement, representatives of NASA and Lockheed Martin met at the Skunk Works – Lockheed Martin's once-secret engineering base in the Mojave Desert. Within thirty months, Lockheed Martin would have to build a half-scale prototype for testing in the atmosphere. They would receive $941 million from NASA, and would have to contribute $220 million of their own. Despite these huge sums of money, Lockheed would have their work cut out to meet the deadline. Paul Landry, chief engineer on Lockheed Martin's X-33 programme, spells out the nature and magnitude of the task: 'The difficulty of putting man on the Moon – to me, it was easier to do that than it is to do this programme; here we have a very aggressive schedule, a cost factor that is limiting us on one side, and the technologies that we're having to try to use to get there.'

It was the very asset that won Lockheed Martin the contract – the necessity of developing new technologies – that would make the project extremely challenging. The team would have to design a new type of airframe that would have to be both strong and incredibly light, and would have to withstand the extreme heat of re-entry. To achieve this, the heat shield will be integrated into the airframe. Unlike the Space Shuttle orbiter's heat shield, comprising thousands of glued-on ceramic tiles, the X-33 – and eventually the

VentureStar – will use nickel alloy panels that are attached to the cage-like bracketing system to form the fuselage itself. While the Space Shuttle's tiles are subject to frequent and detailed inspection, re-waterproofing and general maintenance, these panels will need only one waterproofing and can be replaced easily thanks to the bracketing system. Weight is a crucial factor, and is fundamental in every aspect of the design. The fuel – liquid hydrogen and oxygen – will be held in huge, state-of-the-art cryogenic tanks that are extremely light given their size and strength. There are two hydrogen tanks, which take up much of the volume of the craft, and one oxygen tank. The oxygen tank is made of aluminium, and is therefore very light for its size. A further weight saving is achieved by attaching the X-33's landing gear directly to the tanks – to the oxygen tank at the front and the hydrogen tanks at the rear.

Another significant weight-saving aspect of the design is to be found in the engines. The VentureStar's aerospike engines have only about one-quarter of the size and weight of a conventional rocket engine that produces the same thrust. Much of the saving in weight is due to the fact that an aerospike engine does not have a heavy, bell-shaped nozzle like other rocket engines. The familiar bell shape of the rocket engine nozzle is designed to confine the plume of exhaust gases to a concentrated stream. The shape of the exhaust plume does not change as the atmospheric pressure changes with altitude, and so conventional rocket engines work at maximum efficiency only at a particular altitude. In an aerospike engine, the exhaust gas – water that is produced by the reaction of hydrogen and oxygen – is pushed out from the sides of the reaction chamber, along a tapered 'nozzle ramp'. It is as though the nozzle has been turned inside out; the result of this is that the outside of the chamber is open to the atmosphere. And so, as the atmospheric pressure changes while the X-33 climbs higher, the shape of the exhaust plume automatically adjusts to maintain maximum performance.

The simple but fiendishly clever design concept of the aerospike has another advantage over conventional rocket engines. Traditionally, steering a rocket engine has been a complicated task, involving tilting the whole engine assembly on gimbals (pivoting components). In an aerospike engine, there are no gimbals – in fact there are no moving parts at all. The reaction takes place either side of the nozzle ramp, and increasing the flow of fuel to one side means that the engine's thrust is unbalanced, forcing the X-33 to which it is attached up or down. Turning – banking or rolling – is achieved by adjusting the orientation of the fins.

The concept behind the aerospike engine was first developed during the 1960s, and was considered for the Space Shuttle. However, the technology remains untested in spaceflight. In October 1997 aerospike engines underwent their first tests in flight in the atmosphere, as part of a one-tenth-scale model of the X-33 attached to a supersonic jet aeroplane, the SR-71, or 'Blackbird' – a spy-plane built at the Skunk Works, by the then Lockheed Corporation, in the 1960s. The SR-71 is the fastest production-model aeroplane ever built. It can fly at speeds of more than 3,200 kilometres per hour – more than three times the speed of sound – and at altitudes higher than twenty-six kilometres. In December 1999 the X-33's engines were tested successfully at full power for the first time, this time in a ground-based test facility.

Both the VentureStar and its prototype, the X-33, will be piloted robotically, by a sophisticated system of avionics (electronics that control flight). The X-33 is half the size, one-ninth of the weight and one-quarter of the cost of the

VentureStar. Overall, the VentureStar will be considerably shorter than the Space Shuttle orbiter – 39 metres long compared with 56 metres. Despite the fact that the VentureStar will be able to carry the same mass of payload, its total mass at lift-off will be considerably less than the Space Shuttle: 1,190 tonnes compared with 2,045 tonnes, just over half the weight. The overall aim of the X-33 programme is to reduce the cost of putting objects into orbit – from the current figure of $22,000 per kilogram down to $2,200 per kilogram).

And so, soon it will be a wingless, penguin-shaped, reusable spaceplane that you will see instead of the Space Shuttle. VentureStar should be operational by 2004, and will make routine flights into orbit from then on. Even if these aims are achieved, this spacecraft will not be the 'space Volkswagen' that many people are looking for. It is designed to carry pre-prepared payloads – satellites or crew for the International Space Station, for example – which will be loaded into standardized modules that slot into the craft's cargo bay. For all its amazing technological advances, it will not be suitable for carrying fare-paying tourists to space hotels. Another project that is under way – the X-34 – is similarly aimed at cargo-carrying flights. The X-34 is a smaller-scale project than the X-33, and the result will be another robot-piloted, reusable craft. It will be launched in the air, after receiving a piggy-back from an aeroplane, but will not reach orbit as the VentureStar will. The X-34 is really just a test-bed: a way of gauging the performance of certain new technologies. Some people think that the X-33 and even the VentureStar have similar goals. Ex-NASA deputy administrator Hans Mark reckons, 'We are not at the point where the technology is ready to commit to a single-stage-to-orbit vehicle. That's why the X-33 which was picked out of these three concepts is really not a vehicle to go into space; it's an experimental aeroplane ... My guess is that we're looking at a twenty-year timescale before we actually have a single-stage-to-orbit vehicle.'

Too long to wait

For most people who want to see affordable access to space, twenty years is far too long to wait. We have watched in wonder the incredible achievements of the American and Russian space programmes. We have seen the pictures of people weightless in orbit; we have seen pictures of people floating free in space, attached only by cables and tubes to their spacecraft; we have seen pictures of the curvature of Earth and the glorious views of the Earth from space – now many people want to experience these things for themselves. The basic technological challenges that make spaceflight different from flight in the atmosphere have been overcome, so people are naturally wondering why space is still not within their grasp. But all we see are more satellites being sent up to orbit by the telecommunications companies and global television empires, and more, but smaller, space probes. And now the most exciting development in spaceflight technology is a robot-piloted vehicle with no facility for taking ordinary people into space. For some people, the pace of development is far too slow.

In an effort to speed things up, an organization based in St Louis, Missouri, set up the X Prize – a competition aimed at stimulating the design and production of just the kind of reusable launch vehicle that is needed to transport passengers into space. The X Prize was announced in May 1996: the chair and president of the X Prize Foundation, Peter Diamandis, explained that the winning spacecraft must 'be able to carry at least three human adults, go to

100 kilometres altitude, which is above the boundary of space, come back safely, and then launch again within two weeks; it needs to be reusable, low-cost and be able to carry you and me into space.' The rules of the X Prize are well thought out, and ensure that the aim of the competition are upheld. They are:

- The entries must be privately funded – this will ensure that governments cannot buy into the prize, which the organizers believe would reduce the chances of the competitive spirit they hope will encourage space tourism.
- The winning craft must reach 100 kilometres, the recognized boundary of space – any higher and sophisticated heat-shielding would be required, increasing costs.
- The winning craft must be able to carry at least three people – this will immediately open up the possibilities for fare-paying would-be astronauts.
- The same vehicle must be reusable within two weeks – this will mean that only a minimum amount of work can be carried out to prepare the craft for its next launch, which in turn should bring down the costs to the potential fare-paying customers.

The X prize competition is based on the aviation prizes that were held from the very earliest days of flight. One of the first was the £1,000 offered by the *Daily Mail* that led Louis Blériot to become the first person ever to fly across the English Channel, on 25 July 1909. Another prize, also sponsored by the *Daily Mail*, encouraged John Alcock and Arthur Whitten Brown to make the first transatlantic flight, an arduous sixteen-hour journey from Newfoundland to Ireland. The two pilots received £10,000 for their efforts. On 21 May 1927 Charles Lindbergh won $25,000 as he became the first person to fly non-stop between New York and Paris. The amount of money offered in the aviation prizes increased over time and with the magnitude of the task. However, they have one important thing in common: the amount of money spent by people attempting to win the prize far outstripped the actual prize money. In the case of Lindbergh, for example, sixteen other attempts were made to achieve the task, and about $400,000 was spent in total.

So offering prizes is a very shrewd way of attracting large amounts of funding to encourage rapid advances in technology. The effect of the aviation prizes was to increase the safety, efficiency and range of aeroplanes, and this in turn made aeroplane flight within the reach of many more people. The advent of affordable air travel would probably have been considerably delayed had it not been for these initiatives. The organizers of the X Prize hoped that it would do the same for space travel, and are offering $10 million to the first team who can achieve the objectives they have set out. The founders of the X Prize include Byron Lichtenberg – who has flown on two Space Shuttle missions – and Erik Lindbergh, the grandson of Charles Lindbergh. At the time of writing, there are seventeen teams registered for the X Prize. All the entrants have expertise in aircraft or spacecraft design. Twelve of the entrants are from the USA, one is from Argentina and one from Russia. There are also three British entrants, though one of them is based in Stuttgart, Germany.

One of the X Prize entrants is Mitchell Burnside Clapp, who was involved in the DC-X programme. Burnside Clapp is unconvinced about the safety and viability of powered vertical landing, like that used by the DC-X. Instead, he prefers horizontal take-off and landing – 'the way people were supposed to fly'. His company – Pioneer Rocketplane

Incorporated – is designing a spaceplane that will take off from a runway under jet power, and will be given liquid oxygen for the rocket engines in-flight by a tanker aeroplane, through an interconnecting hose. The runway and tanker refuelling operation are existing technologies, making this approach to reaching space routine and achievable. 'We are attempting to adopt an aviation pattern from the beginning,' explains Burnside Clapp. The jet and rocket engines aboard the Pathfinder will be powered by kerosene. The tanker aeroplane will do the hard work of carrying the liquid oxygen to altitude. Transferring the liquid oxygen in the air should also make the launch site safer. At 110 kilometres, the Pathfinder's payload bay doors will open, and the payload can be launched into orbit. The spacecraft will then glide back to Earth in the same way as the Space Shuttle orbiter does, but the jet engines fire again to help control the craft during landing. Because this spaceplane is similar to a conventional aeroplane, and is piloted, it could even travel to a designated landing site to pick up a satellite which it will then go on to launch into orbit. Apart from the transfer of liquid oxygen from a tanker aeroplane, this whole system depends upon tried-and-tested components and techniques.

As with many of the X Prize entrants, Pioneer Rocketplane was not formed simply to compete for the $10 million prize. As their product is composed of existing technologies, they believe that they can offer significant reductions in the cost of launching satellites within a few years. This idea was given support in 1997, when Pioneer Rocketplane was among four companies awarded a contract to develop a 'bantamweight satellite' launcher. Such a satellite has a mass of about 150 kilograms.

Some of the other entries to the X Prize will also rely on conventional aeroplanes to help them on their way into space. For example, the Russian entry, Cosmopolis 21 – designed by a company of the same name – will be a space-plane that sits upon a powerful carrier aeroplane to an altitude of 20 kilometres. At this height, the carrier aero-plane will turn upwards, into a steep ascent, and the spaceplane's rocket engine will take over, launching the spaceplane above the boundary of space. Another entrant that will use a conventional aeroplane to help it into space is the Eclipse Astroliner, proposed by Kelly Space and Technology Incorporated, based in San Bernadino, California. The Astroliner will be a delta-winged craft that will be towed into the air by a Boeing-747 jumbo jet, rather like the way a glider is launched. The company claims that towing the spaceplane into orbit is safer than attaching it directly to the fuselage of an aeroplane. The Eclipse Astroliner will then be released, and will use a conventional kerosene-powered rocket engine to fly to 182 kilometres, deposit its payload and glide back to Earth.

Another company competing for the X Prize is Scaled Composites, based in the Mojave Desert, near to Lockheed Martin's Skunk Works. They also propose giving its space-craft a piggy-back on a conventional aeroplane. Their entry is Proteus, a spaceplane that will be taken on a specially designed aeroplane up to 11 kilometres, and from there use rocket power to thrust up to the 100-kilometre point stipu-lated by the X Prize. The aeroplane part of the design has already been built, tested and exhibited at several air shows. It is made from an ultra-light composite material – the speciality of Scaled Composites, which is normally more concerned with aeroplanes than with spacecraft. The com-pany was founded in 1982 by Burt Rutan, an ex-air force engineer with a great deal of experience in the aerospace industry. On 14 December 1986 Rutan and Jeana Yeager took off from Edwards Air Force Base, also in the Mojave Desert, in a remarkable aeroplane – called Voyager – which Rutan had designed. It carried just 6,000 litres of fuel, but nine days later it arrived back at Edwards Air Force Base, after flying around the world without refuelling. What made this possible was a remarkable composite material designed by Rutan himself. Voyager weighed less (when empty) than a car, despite its 33-metre wingspan. The first stage of Rutan's X Prize entry – the aeroplane – is made from the same material as Voyager, but the second, rocket-powered spaceplane stage is still under wraps. Rutan is determined to make his dream of affordable spaceflight a reality. He is unhappy with the approach taken by NASA: 'They are in the way, instead of helping us, as far as you and I going into space is concerned ... it is as if they don't want us to ever go.'

Most of the entrants to the X Prize are spaceplanes, though not all of them are released by conventional aircraft. The Cosmos Mariner, for example, designed by Dynamica Research, is among those that will take off from a conventional runway, carrying everything it needs to make it into space. The main problem to be overcome by single-stage-to-orbit spacecraft concerns weight: at take-off, a rocket must carry all its fuel and an oxidizer (normally liquid oxygen), in huge quantities, and this greatly increases the weight. The Pathfinder, described above, gets around this problem by taking on oxygen from a tanker aeroplane at altitude. Others, such as Rutan's Proteus system, will carry the spacegoing vehicle up to altitude, so that it does not need to carry so much fuel. The approach taken by Dynamica Research is different: it will use an 'air-breathing' jet engine while in the atmosphere, and then switch to rocket power when the air becomes too rarefied – just like

Alan Bond's HOTOL would have done. The rocket engine will be switched on at a height of about twelve kilometres, as a result of which the Cosmos Mariner will shoot up to above 100 kilometres, to glide back down to Earth again. The same approach is taken by one of the British entrants – Bristol Spaceplanes, based in Bristol – with their space-plane, called Ascender. Similar again is the X Van, another jet-and-rocket spaceplane, built by Pan-Euro Inc. There is one significant difference with the X Van: the launch will be vertical, like a rocket, and the first phase of the flight – to 10 kilometres – will be near-vertical. The craft will glide to Earth, like the other spaceplanes.

While some entrants to the X Prize intend to use rocket power only, and others intend to use a mixture of jet power and rocket power, there is one who intends to use jet power and no rocket power. He is John Bloomer, of the Discraft Corporation. His proposed spacecraft is called the Space Tourist. Because the jet engines need air to operate, they will be used only below about sixty kilometres. The Space Tourist's powerful pulsed-jet engines will accelerate it to a high enough speed to shoot it above 120 kilometres. The fact that the X Prize rules stipulate only that the craft reaches space does not mean that the winning craft must go into orbit. And so this ballistic approach – one big thrust, then let momentum and gravity do the rest – is acceptable. Indeed, none of the entries is designed to go into orbit. One way or another, they are all designed to travel to extremely high altitude, where gravity will pull them back down again.

So far, we have only heard about spaceplanes. Some other entrants to the X Prize propose using a more tried-and-tested way of getting into space: vertical take-off, missile-shaped rockets. One of the entrants taking a chance on traditional rocketry is Dr Graham Dorrington, based at

Flight Exploration in Stuttgart. His design, Green Arrow, will use kerosene fuel, but with hydrogen peroxide rather than liquid oxygen. All the fuel will be used to produce one powerful thrust, and the spacecraft will coast up to 100 kilometres, from where it will return to have its descent slowed by parachute and its landing softened by gas-filled balloons. Other designs will use rockets like this and various parachutes, balloons or aero-shields, to slow their descent. Thunderbird, put forward by Steven Bennett of the Starchaser Foundation, based in Dukinfield just outside Manchester, is one of them. Thunderbird looks like a conventional, missile-shaped rocket but, like many of the other entries, it will use a combination of jet and rocket power. Bennett comments, 'You don't have to be a rocket engineer to do this. The technology to do what we're proposing to do has been around for the past forty years ... It's just a case of bringing these existing things together in a smart way and bring them together in a unique way to crack the problem.'

The most unusual entry to the X Prize is Roton – a space helicopter being put forward by HMX Incorporated. Roton looks like the DC-X – a huge, upturned cone – but it has one important difference. At lift-off, Roton uses a rocket engine at its base, just like the DC-X. It climbs vertically, burning the kerosene fuel and liquid oxygen. For its descent, Roton will deploy its rotor blades, which spin like the blades of a sycamore seed falling slowly through the air. This will provide Roton with a controlled approach to landing. Burt Rutan, whose company Scaled Composites is producing the material from which Roton is being made, believes that one of the good points about Roton is its safe descent: 'I'm a helicopter pilot, and I know that you can take a helicopter to any altitude and fail the engine, and it's actually a very good way to get down.' The landing will be controlled yet further by firing up the rotor blades, which will have small rocket engines at their tips. In the original form of the design, the powered rotor blades actually provided lift to raise the craft off the launch pad, and to help with the first phase of ascent. The spinning action of the rotor blades would be used to pump fuel at high pressure to the main engines, situated in a ring around the base of the spacecraft, increasing the engine's efficiency dramatically. However, to reduce the financial risks involved in the project, development work on this design has been deferred, and a conventional rocket engine will now be used.

The concept of Roton was devised by Bevin McKinney, who describes it as 'not anything like an existing rocket configuration'. McKinney's colleague Gary Hudson remarks: 'Usually after ten minutes of saying, "Gee, that can't work", they say, "Hmm, actually that helps you here, or here", and by the time they're done they're asking if we need another consultant.' Whether it wins the X Prize or not, Roton may be here to stay: the atmospheric test version of the Roton was 'rolled out' in March 1999, when it hovered its way through a range of tests. A company set up to take on the Roton concept – the Rotary Rocket Company – hopes to conduct orbital tests early in 2001, and to go into commercial service later that year.

In addition to Roton, some of the other vehicles competing for the X Prize will also still be built if they lose the competition, and will probably go into service. In addition to the X Prize, there are several other spaceflight awards on offer – with smaller prizes, but just as much prestige. Examples are the CATS Prize (Cheap Access to Space) and the FINDS Prize (Foundation for the International Non-governmental Development of Space). These will help to ensure that the human race quickens its step in the march to space. In fact, the pace is quickening quite naturally – the prizes are just a catalyst, to encourage the march to become a jog or even a sprint.

As well as the governmental and private rocket-building projects, there is a growing number of private individuals or groups investing time and money into their own space race. Some are simply weekend rocket enthusiasts – hobbyists out to dream of the day when they may leave the Earth. Others are taking part in the private space race, experiencing first hand the thrill of launching something they have made so high that it comes close to the boundary of space. And in the past few years, some have come tantalizingly close. In 1998 a group succeeded in launching their home-built rocket so high that the on-board video camera could capture a stunning view of the Earth. The flight controller was ecstatic: 'Oh my God, the curvature of Earth ... We've got it on video ... I can see the curvature of the ... Look at that, look at that! It's the friggin' Earth, man! That is so cool!' The sense of excitement is clear. One approach taken by the amateurs, which overcomes the problem of increased air resistance at lower altitudes, is to use helium balloons to lift their rockets up into the rarefied atmosphere before launching them. This idea, borrowed from the 1950s, has allowed one team to set an altitude record of 21 kilometres.

There are people who believe that the pace of the human race's march to space should be slowed rather than speeded up. They are not many in number, and their voices are not loud over the excited murmur of the space enthusiasts, both amateur and professional. But there are plenty of reasons why the drive to increase space traffic may be something we need to think about. There is environmental concern over noise and pollution of the upper atmosphere. There is the use of huge amounts of natural resources, and the amounts used will continue to increase as space becomes more and more accessible. And then of course there is the fact that space projects will benefit only the privileged few.

There are other people whose voices are in dissent against the majority: those who think that there is nothing wrong with the human desire to explore space – just with the assumption that it might happen cheaply and very soon. Dr Hans Mark is one of them: 'I've always worried about the term "cheap access to space"; I think we're fooling ourselves. To get into space, you have to achieve an orbital velocity of 17,000 miles per hour, and that's never going to be cheap.' He believes, too, that competitions like the X Prize are irresponsible. He says that the risks involved in air travel are small and well known, but, in space travel, the risks are simply too high. Burt Rutan believes otherwise: that 'if no one dies going after this prize then we are not really going out truly searching for new ideas'. For him, risk is a necessary part of progress: 'If we had shut down things because there's risk being taken, we would still be travelling across country looking up the assholes of donkeys; that's not the future and that's no fun.'

Whatever the concerns of dissenting voices like Hans Mark's, or those who are concerned about space travel for environmental reasons, private enterprise will almost certainly move quickly into a future in which space travel really is relatively affordable and routine. As well as the space hotels or joy-rides into space, there are possibilities that cheap access to space could one day replace long-haul air travel. Michael Wallis is a space enthusiast and entrepreneur. He has a clear view of the future: 'When you're going through space to get somewhere, you can get anywhere on the surface of the Earth in forty-five minutes. It means that London, for example, is thirty-one minutes from California … You can travel at speeds because you're moving out of the atmosphere.' He also sees great opportunities for transporting cargo, say from Japan or New Zealand: 'Pick up by nine in the morning; delivery to the United States or Canada by 5 pm the previous day – for those occasions when it really did have to be there yesterday.'

Burt Rutan also believes that cheap and reliable access to space will one day replace air travel. 'To sit in an airliner burning fuel continuously at 30,000 feet going to London can't be the best way to go there. It would be a lot more fun – and we're going to find very affordable ways to do this – to make a big-time thrust for a few seconds and then to sit there and enjoy your meal in weightlessness and have a much better view, and then re-enter over London and land. That's got to be a better way to travel.'

WHAT SHALL WE DO WITH THE MOON?
... searching for a future for the human race...

The Moon has been our companion for about 4,500 million years, silently circling our planet in the dark vacuum of space. With its mysterious beauty, the Moon has been the subject of myth and wonderment for millennia, inspiring poets and scientists alike with imaginative ideas. But in the past hundred years or so it has captured people's imaginations in a new way: the advent of space travel has brought new fanciful notions, such as the idea that humans may one day colonize the Moon, setting up permanent bases or even lunar cities. This idea may not actually be fanciful, and the day it is realized may come along sooner than you think. And once the bases have been built, the Moon could be just the first stop-off point for journeys much farther afield.

Bright future
Perhaps the most obvious and magical thing about the Moon is its phases: the fact that it changes its appearance gradually over a period of a month, from a thin crescent to a full bright white disc and back again. The phases are caused by the fact that the Sun illuminates half the Moon. As the Moon orbits Earth, once every month, we see different amounts of the illuminated face. When the Moon is between the Earth and the Sun, we cannot see the illuminated side at

all – this stage is called new moon. When the Moon is pre-cisely in line with the Sun – something that does not occur every month because the Moon's orbit is tilted by about five degrees to the Earth's orbit around the Sun – the Moon blocks out the Sun, and we have a solar eclipse. This happens only once every few years, and only at new moon. The day after new moon, when the Moon has moved on in its orbit, a thin crescent of the illuminated face of the Moon is visible from Earth. Just to confuse things, this crescent is sometimes called a new moon.

When the Sun and the Moon are at right angles to each other as seen from Earth, half the Moon's illuminated side is in view. This is a half moon – although this phase of the cycle is called quarter moon, because it occurs when the Moon is one-quarter or three-quarters of the way around its orbit. When the Moon is on the opposite side of the Earth from the Sun, we see its entire illuminated face: the full moon. Again, because of the tilt of the Moon's orbit it is rare for the Earth, Sun and Moon to line up. When they do, we have an eclipse – this time a lunar eclipse, where the shadow cast by the Earth falls on the Moon.

Although ancient astronomers had no idea what the Moon actually was, they were able to observe and then pre-dict its movement across the sky and the way its phases changed. They were also able to predict when lunar and solar eclipses would occur – a task that involves complic-ated mathematics.

The silent, startling beauty of the partially or fully vis-ible Moon dominates the sky on a clear night. The Moon is the second brightest object in the sky, after the Sun. The main reason for this is its relative proximity to Earth: the Moon is our nearest celestial neighbour. In fact, the dis-tance between the Earth and the Moon is, on average, only 384,400 kilometres – less than ten times around the world. And, unlike travel around the world, in space there is no air and therefore no air resistance. So once you are outside the atmosphere, in orbit around the Earth, it is relatively easy to give yourself the thrust necessary to leave that orbit and head for the Moon. Then the Moon is only about three or four days' journey away. If your thrust was just right, you would move into an orbit around the Moon instead of around the Earth, and from there it would be easy to use your rockets to slow yourself down and descend to the lunar surface. What would you find when you got there?

First, you would have a surface area about four times that of the USA to explore. The Moon's diameter is 3,476 kilometres – a little less than the distance between New York and Los Angeles. Imagine slicing the Moon in half along its equator: one of the resulting hemispheres would cover most of the USA. Although the Moon looks bright, it produces no light of its own and reflects only about seven per cent of the sunlight that hits it. If it were painted white, it would reflect nearly all the incident sunlight, and it would then appear almost as bright as the Sun itself. But the Moon's surface is actually dark grey, so it reflects only that small proportion of sunlight. As most of the sunlight is absorbed, most of its energy is also taken in. This in turn means that the illuminated surface gets very hot. Each point on the Moon is repeatedly in constant sunlight for two weeks and then constant darkness for two weeks. So during the long lunar 'day', the temperature of the surface becomes very hot, reaching 130 degrees C. During the lunar 'nights' the surface re-radiates most of the energy into space, as infrared, and its temperature drops to a more than chilly minus 110 degrees C.

The length of the lunar day is a result of the Moon's slow rotation on its axis. The Sun moves very slowly across the lunar sky, taking two weeks to cross from the eastern horizon to the western horizon. The sky remains dark even when the Sun is 'up', since there is no atmosphere on the Moon to scatter the Sun's light and create daylight. The Earth also drifts slowly across the lunar sky when it is in view. It appears nearly four times as large in the sky as the Moon does from Earth.

The gravitational attraction pulling you down on to the Moon's surface would be only about one-sixth of the gravitational force pulling you on to the Earth's surface. This is why astronauts who go there can bounce along, taking giant steps, and can carry heavy objects with ease. And whereas you cannot easily jump more than about thirty centimetres on Earth, you would be able to push yourself several metres off the surface of the Moon, and fall back slowly and gracefully. If there was a smooth surface, like a polished floor, you would have trouble walking along normally – because the traction you need to start walking comes from the friction between the floor and you. That friction depends upon how hard gravity pulls you down on to the floor, and so on the Moon your feet would slip and slide at first.

The lack of atmosphere on the Moon has several fascinating consequences, too. Many of the things we take for granted here on Earth are radically different on the airless Moon. Striking a match would produce a brief burst of flame, as the chemicals in the match tip ignite, but the wood would not burn. A balloon pump would be easy to use, as there would be no air pressure inside it to resist your pushing on the plunger. However, for the same reason, the balloon would not inflate. Anything that relies on aerodynamic surfaces – such as an aeroplane – would not work on the Moon as it does on Earth. No lift force would be generated by the wings, however fast you thrust the aeroplane forward. If you dropped a feather and a hammer simultaneously from the same height, they would reach the ground together, because with no air there is no air resistance. This was demonstrated by Apollo astronauts. Furthermore, sound needs substance through which to travel, and so you could shout as loudly as you like through the visor of your spacesuit, but a friend standing right next to you would hear nothing.

The effects on people are perhaps the most dramatic. If you went to the Moon, you would not be able to survive unless you were in a pressurized container such as a spacesuit, and were supplied with oxygen. If you removed your spacesuit, your eyes would bulge, the air in your lungs would immediately rush out into space, and your lungs would collapse. You would quickly die. One other thing: the Moon offers ideal conditions for observing the stars and planets. On Earth, the best views of the starry night sky are seen on calm, clear, dark nights. Even then, the sea of air through which you have to view the stars is in constant motion – pockets of warm air rise and mix with cold air, for example. When you look at the sky from the Moon, light coming from the stars and the planets does not have to pass through a turbulent atmosphere as it does on Earth. And there would never be any clouds to obscure the view.

Making an impact

Most of the surface of the Moon is covered with regolith, a mixture of rocks and sand, which is formed when large meteoroids hit the Moon. These heavy, fast-moving, interplanetary rocks make craters when they collide with the Moon, shooting thousands of tonnes of material out in all directions. The shockwaves created in a powerful impact heat the surface material, which melts then quickly cools and solidifies, forming tiny glassy globules. The existence of these globules in the regolith was hypothesized before people or space probes made it to the Moon – and it was confirmed by analysing samples of regolith brought back by the Apollo astronauts and by Russian robot probes.

Regolith covers most of the lunar surface because craters have been formed all over the Moon. Some areas do have fewer craters than others, however: these are the 'maria' (plural of 'mare', Latin for 'sea'). There is no water in these seas: they are flat areas of solidified lava that has filled ancient craters. The maria are the dark areas you can see with the naked eye when you look at the Moon in the sky. Maria are less well populated with craters because they are relatively young – most of the Moon's craters are very old. This in turn is because meteoroids struck with much greater frequency in the Moon's younger days, when the Solar System was bustling with the remnants of its own formation. It was at this time – theory has it – that the Moon was formed, from a huge cloud of material thrown into orbit around the Earth. That cloud of material was ejected from Earth when a huge object collided with our planet in the very earliest period of the Solar System – about 4,600 million years ago. Most of the rocks on the Moon, then, date from this period – much older than the rocks found on Earth.

Our planet is a changing place: weathering constantly changes the chemical composition of the Earth's rocks. And new rocks are made, or existing rocks changed, by the movement of the tectonic plates of its crust. For these reasons, there are almost no rocks older than about 3,000 million years left to study on Earth. But the Moon has no tectonic plates, and no weather. Over time, meteorite impacts have changed the physical properties of some of the rocks – producing regolith – but the chemical composition of the rocks has not changed. So the Moon provides a good picture of the chemical composition of the early Solar System.

The craters that form when the Moon is hit by heavy meteoroids are a familiar feature of the lunar surface. The very biggest ones are visible to the naked eye from Earth, because the material that is thrown out by the impact produces large streaks that radiate out in the surrounding areas. When you look at the Moon, one of the most prominent craters you can see is Copernicus, named after the sixteenth-century Polish astronomer Nicolaus Copernicus. At its edges, this crater has terraced walls, which form huge steps that may one day allow people to make their way down the 3.8 kilometres from the rim to the base of the crater, which measures 93 kilometres across. To the right of Copernicus on the face of the full moon is a relatively flat, uncratered area called Mare Tranquillitatis – the Sea of Tranquillity – where the first human being stepped on to the Moon, in 1969. It is also the first place from which rock was collected on the Moon. Before that date, only robotic probes had actually landed on the surface, and none had returned any lunar material.

Throughout the 1960s a series of Russian and American probes flew by, orbited, hit or landed on the Moon. None of these probes carried any people. It was Apollo 8, in December 1968, which first took people to the Moon. The astronauts aboard the Apollo 8 spacecraft made ten lunar orbits before returning home. As became typical of the Apollo programme, the people still down on Earth could monitor the progress of the astronauts in a series of live television broadcasts. During their Christmas Eve broadcast – from lunar orbit – the crew of Apollo 8 read sections of Genesis from the Bible, while showing the people of Earth the beautiful views of their planet and the Moon from space. The pilot of the command module, Jim Lovell, said: 'The vast loneliness is awe-inspiring, and it makes you realize just what you have back there on Earth.' It is no surprise that the Apollo missions inspired many people.

Apollo 9, launched in March 1969, did not go to the Moon. Instead, it went into orbit around Earth, to test the operation of the lunar module, identical to the one that would descend to an orbit close to the lunar surface in the next mission. Apollo 10 lifted off in May 1969, and completed this mission successfully. Two months later, Apollo 11 took Neil Armstrong, Edwin 'Buzz' Aldrin and Michael Collins to the Moon. And so on 20 July 1969 Armstrong became the first person ever to step on to the Moon. It is the fact that only twelve people have done this so far – despite more than thirty years of endeavour – that leaves Moon enthusiasts frustrated and eager to go to the Moon themselves.

Including the regolith they collected, astronauts and space probes have so far brought 382 kilograms of material back with them from the Moon. Some of the minerals available on the Moon could be of economic importance: there are huge amounts of aluminium, titanium, magnesium and iron there, for example, and silicates, from which you can make glass. Pure silicon can be extracted from silicates to make electronic components, which could consist of perfect crystals if they were manufactured in the weightless conditions of a Moon-orbiting factory. The metals could be extracted from the rocks, using the heating effect of sunlight or electrical smelting plants running on solar power. Products of these lunar industries could be used to build space stations or spacecraft far more cheaply than on Earth.

This is because lifting thousands of tonnes of metal off the surface of the Moon requires a lot less thrust than doing the same from the surface of the Earth, and launch – as we have seen in 'Day Return to Space' – is the most costly part of space travel. This is easy to appreciate when you compare the size of the Apollo lunar modules that blasted off from the Moon with the size of the Saturn V rockets used to blast the mission off from Earth in the first place. The difference is in the force of gravity against which the engines were fighting.

There is another valuable resource on the Moon: an isotope of helium, which may be the fuel for nuclear fusion power stations. This could be of importance in the coming century, as the population and its energy demands increase. Fossil fuels are used to generate electricity and to power most forms of transport, but burning them releases pollutants, including millions of tonnes of carbon dioxide, into the atmosphere every year. The use of helium-3 as the basis of fusion, to generate electricity for the power-hungry modern world, would reduce the need for fossil fuels.

Mining and lunar space-vehicle construction may not be the only technological projects to be staged on the Moon. In another effort aimed at reducing our dependence on fossil fuels, Dave Criswell, of Space Systems Operations at the University of Houston, has proposed a way of using solar power collected on the Moon here on Earth. Criswell's idea is to place huge arrays of solar cells on to the Moon's surface, to change sunlight into electrical power. There are no clouds on the Moon to obscure the Sun, and so the cells would provide a reliable source of electrical power for as long as they are in the light. Having solar cells in only one part of the Moon would mean that power would be generated for only two weeks out of every month – when that part is illuminated. So Criswell proposes covering vast tracts of the lunar surface with solar cells, ensuring a constant supply. The electrical energy would be used to produce high-power microwave radiation that would be beamed to Earth, where the energy could be changed back into electricity.

Criswell summarizes the project: 'Solar power is the ultimate resource in the Solar System ... so the challenge is to divert a tiny bit of it efficiently down here to Earth.' But if a massive mining operation or Criswell's lunar–solar power scheme does go ahead, people will need to live on the Moon for extended periods of time. Moon bases will have to be built. And if that happens, Criswell compares what would happen to the Moon as something like the colonization of mining towns in Colorado in the nineteenth century: 'People will find reasons to be there, most of which have nothing to do with the primary activity.'

Despite these thought-provoking possibilities, to nearly all the world the exploration of the Moon seems to have halted. The last human beings to stand on the Moon were Eugene (Gene) Cernan and Harrison Schmitt, during the Apollo 17 mission in December 1972. The Apollo programme was the jewel in the crown of the USA's efforts in space: twelve of its astronauts stood on the Moon, collected samples and even drove around on the surface in a remarkable lunar rover. After the Apollo programme finished, many people assumed that visits to the Moon would continue – NASA themselves had plans for an exciting future. Dr Alan Binder of the Lunar Research Institute in Gilroy, California, recounts this lost opportunity: 'We had a post-Apollo programme planned which would have put a lunar base up by the end of the 1970s, and there would have been fifty to a hundred people living and working on the Moon by the early 1980s; we simply dropped that.' And so, since 1972, only robotic probes have been back to the Moon.

After the flurry of lunar exploration during the 1950s, 1960s and early 1970s, planetary scientists moved on to explore the deeper realms of the Solar System, and probes have now also visited all the other planets except Pluto. You would be forgiven for believing that we have forgotten the Moon in favour of more exciting destinations. But the dreams of building permanent bases on the Moon – so much in currency during the Apollo era – have remained alive, and have recently undergone a renaissance. One reason for this resurgence in interest in the Moon is that the people who were inspired by the Apollo missions are becoming impatient.

Patrick Collins of the Space Development Agency of Japan is one of the impatient ones: 'People went to the Moon thirty years ago; now if you do something thirty years ago, that means it is easy today.' Another person who is very keen for humans to set themselves up on the Moon is aerospace engineer Greg Bennett – Director of the Lunar Resources Company, based in Houston, Texas. He says, 'If you were to go to the Moon thirty years from now, you would think that the Moon has simply become the Hong Kong of the Solar System.' Space capitalist Jim Benson is another impatient lunar enthusiast: 'I've been waiting for something to happen in space for most of my life … I'm getting tired of waiting.' Benson runs SpaceDev, one of a number of companies already making millions of dollars from the prospects of lunar exploration. He and his shareholders are speculating on the bright future for projects that exploit the potential of space. He believes that 'space is too expensive to do as a taxpayer-supported programme; we need to find ways to make space pay and then we can do anything that makes a profit, we can go to space and do all those things'.

The obvious things that people could do to make money from the Moon are mining and tourism. And when people start organizing such activities, Benson will be there, to make money: 'When I think of lunar colonies, I think of all the business opportunities ... If SpaceDev can supply the infrastructure, it's going to be a very profitable company.' So confident is Benson of the impending moon rush that he has built new offices in Poway, California, which include a clean room where spacecraft will actually be built and serviced. Benson's company deals with all aspects of the commercialization of space – including proposed mining on asteroids – but the Moon has special significance, as it could become the launch pad for journeys to more far-flung locations. The offices also include a 'mission control' room from which personnel will monitor and control ongoing missions. Benson explains that there will be 'four or five personal computers and monitor screens' inside this unlikely-looking hub of activity. SpaceDev's first project – the Near-Earth Asteroid Prospector that plans to mine the resources of asteroids – was begun in 1997. This is a significant year in the history of space exploration: it was the first year in which global commercial expenditure in space projects outstripped government spending. This situation is likely never to be reversed, and in fact the gap between publicly and privately funded space projects will probably only widen. The Moon is ripe for development and exploitation and, like it or not, the commercialization of space seems to be here to stay.

Don't go there

There are many obstacles and objections to the construction of the first Moon bases and cities. One of the principal objections concerns the economics behind such a programme:

the money spent in setting up a base on the Moon – whether from public or private sources – could be used to feed the millions of starving people here on Earth. If the pattern of Earth-based industry is followed on the Moon, then the economic benefits of lunar exploitation are unlikely to be shared among all people throughout the world. However, many would argue that while industry makes only some people rich, it does bring incalculable benefits to everyone, by developing new technologies or manufacturing goods that then become available to everyone. To keep in check the 'get-rich-quick' mentality that could turn pioneering lunar exploration into a mere money-making exercise, a United Nations treaty was created in 1967, to which several of the industrial nations signed up. The 'Treaty on Principles Governing the Activities of States in the Exploration and Use of Outer Space, Including the Moon and Other Celestial Bodies' included the following requirements on signatory nations:

- 'The exploration and use of outer space, including the Moon and other celestial bodies, shall be carried out for the benefit and in the interests of all countries, irrespective of their degree of economic or scientific development, and shall be the province of all mankind.'
- 'Outer space, including the Moon and other celestial bodies, shall be free for exploration and use by all States without discrimination of any kind, on a basis of equality and in accordance with international law, and there shall be free access to all areas of celestial bodies.'
- 'There shall be freedom of scientific investigation in outer space, including the Moon and other celestial bodies, and States shall facilitate and encourage international co-operation in such investigation.'

So much for the regulation of our futures in space, which may restrain some of the overzealous lunar explorers. There are also innumerable technological obstacles to be overcome in the establishment of lunar bases. The first thing that will be needed is an affordable and reliable method of launching people and supplies from Earth, transporting them across space and landing them safely on the Moon. Once there, lunar pioneers would have to survive on an airless ball of rock: the Moon does not have an atmosphere. Here on Earth, the atmosphere not only provides us with air and water: it blocks some of the sunlight that reaches our planet, reducing potentially carcinogenic ultraviolet radiation. The Moon would afford no such protection from the harmful effects of direct sunlight. Earth's atmosphere also keeps the planet's temperature at just the right level for life to thrive. In fact, the composition of the atmosphere is a direct consequence of the existence of the living things that it supports. Without the atmosphere – in particular, the carbon dioxide and water vapour constantly cycled and regulated by living things – the average temperature at the surface would be about a hundred degrees C cooler. So the Moon is not only airless: it is very cold, too.

The astronauts who have already made it to the Moon could not stay for long. This is mainly due to the fact that they had to take all their food with them – with no atmosphere and no known source of water on the Moon, they could grow nothing. Astronauts have stayed in space for extended periods, in orbit around the Earth, but this was possible because spacecraft would bring fresh supplies of oxygen, water and food. The methods used to maintain the right conditions inside orbiting space capsules can do so for only limited periods of time. Exhaled carbon dioxide is removed from the capsule's air by cylinders filled with a substance called lithium hydroxide, but these cylinders must be changed daily. It would not really be viable to send frequent batches of supplies to a base on the Moon designed to accommodate hundreds or thousands of people.

If the requirements of making it to the Moon and staying alive there are so challenging, how can people be so confident that within their lifetimes there will be permanent bases or even civilizations in this hostile environment?

The first requirement for living on the Moon – a reliable method of transportation – is not yet a reality. Despite the fifty years or so that have passed since humans first propelled rockets into space, the dream of routine, affordable space travel seems no closer. But a new entrepreneurial spirit may be about to change that. As we saw in the previous chapter, the space technology developed in the space race of the 1950s and 1960s was not designed to make space travel routine and affordable. The impressive and exciting achievements of the American and Russian space programmes were driven by political goals as much as – or possibly more than – by scientific ones. But initiatives such as the $10 million X Prize may be the stimulus that private space travel engineers need to design and build the kind of reliable vehicles that will eventually take people and supplies to the Moon.

Assuming that we can make it to the Moon, how will the other requirement be met? How will lunar pioneers sustain themselves if there is no atmosphere? This is perhaps more of a challenge. Even if lunar engineers took billions of tonnes of the atmospheric gases from Earth and released them at the Moon's surface, the gas molecules would soon escape from the weak gravitational pull there, and boil away into space. So if people do ever stay for extended periods on the Moon, they will need to be enclosed in an airtight building, breathing air from an artificial atmosphere. If a leak developed in the building, the atmosphere would escape and the lunar inhabitants would die, so the pressure of the building would have to be monitored constantly.

The good life

Surviving on the Moon is not as simple as just living in an airtight building, however. You would have only the precious resources that you had taken with you. So lunar inhabitants would be unwise to throw their waste away on to the Moon or into space, since otherwise they would soon run out of water and raw materials. Everything would have to be recycled meticulously – residents of the Moon would have to be totally self-sufficient. A lunar base would have to be a self-contained mini-ecosystem, in which all of the resources would have to be carefully monitored and controlled. This requirement, too, may not be as much of a challenge as it seems. What is needed is a large-scale version of a terrarium – an enclosed container used to propagate plants or certain animals in carefully controlled conditions. The terrarium was popular as a form of decoration during the nineteenth century, as well as a being a tool for scientists and plant- and animal-keepers.

In several experiments conducted on Earth with extraterrestrial survival in mind, teams of people have survived in the equivalent of large terrariums for extended periods. The first such sealed environment was 'Biosphere 2' – a steel and glass building covering just over 1.2 hectares of land in the Arizona desert. Four men and four women lived sealed inside Biosphere 2 for exactly two years, from September 1991. Some of the scientists involved in the project had space survival in mind, although Biosphere 2 was set up as an experiment to model the features of Biosphere 1 –

the Earth. The building contains 170,000 cubic metres of air, and has areas dedicated to rainforest, desert and farmland. It also has an ocean, containing 3,800,000 litres of salt water. In similar experiments carried out by NASA – this time with survival in space very much in mind – groups of researchers spent weeks inside a sealed unit nicknamed the 'Chamber', at the Johnson Space Center in Houston, Texas. This steel-walled container is 6 metres in diameter and three storeys high. Scientists on the outside monitored the physiological and psychological condition of the volunteers inside, through video links and medical tests. Future self-contained living spaces like the Chamber could be packed into a spacecraft that was going to the Moon.

The Chamber was developed as part of the Lunar–Mars Life Support Test programme, begun in 1995. It was as long ago as the 1950s when research began into the use of plants for human survival, and NASA took an interest during the 1960s. A tank similar to the Chamber has been in use at NASA's Kennedy Space Center in Florida – to study the production of plants in sealed environments – since 1986. During the first experiment in the Chamber, four volunteers spent four weeks inside. The longest stay inside the Chamber began in September 1997, and lasted for thirteen weeks. The team entered the Chamber with enough air and water to last only one week, and had no supply from outside. So inside, everything was recycled. The water they drank and washed in was obtained by recycling their own urine. The carbon dioxide they exhaled, together with carbon dioxide obtained by incinerating their faeces, was made available to plants that were kept in hydroponic and soil-based tanks. (In hydroponic cultivation, plants grow in soil-less conditions, suspended in nutrient solutions.) As the plants used carbon dioxide, they produced oxygen – which the inmates needed to breathe – and sugars, which the plants needed if they were to grow. Once grown, the plants themselves were a source of food, which could be eaten, and as solid and liquid waste, be recycled once again.

This regenerative approach – using the same raw materials again and again – mimics Earth's own natural recycling system. Some of the plants grown in the Chamber were eaten, but the inmates in this experiment ate mostly tinned and frozen food. In space, it would be uneconomical to take much food with you – an adult human being consumes several hundred kilograms of food each year. So any advanced life-support systems would have to have a facility for growing food. The chief scientist at the Johnson Space Center, Dr Don Henninger, explains that inhabitants of a colony on the Moon could grow plants in the lunar soil – from the regolith. Henninger runs experiments in which he grows plants in 'lunar simulant soil … a high-titanium basalt'. In this lunar soil, he has grown specially developed dwarf wheat, which grows only about thirty centimetres high. He has also grown potatoes, soya and even some herbs and spices. So you could actually use the soil from the Moon to grow crops. Of course, as Henninger points out, 'Since there is no atmosphere, we'd have to bring it inside some habitation chamber or plant growth area.'

The team inside the Chamber evaluated a diet that was designed for life on the lunar surface, which included fifteen different crops. One of the inmates in the Chamber was Nigel Packham. Though he survived on the diet well enough for the duration of the experiment, he says, 'I'm not sure if taste-wise you'd want to do that for three, four or five years or longer.' As well as having a restricted menu, the chamber provided a very confined living space. It did have a games area and a library, and there were rooms where

Packham and his colleagues could be alone when they wanted to be. But on the whole it was not very comfortable, and there was not much space to move around. Packham says that his experience gave him 'some idea of what a lunar base might be like'. However, the confinement that the inmates felt living in this enclosed container would have been intensified if that container were up on the Moon – where they would not be able to step outside back into normality. The project's psychologist, Al Holland, says, 'Anything you can do to normalize that life is important, which includes the use of plants, not only for physico-chemical regeneration but also for the psychological regeneration that occurs with contact with plants ... If we could have pets on a lunar colony, if they weren't too distracting or too disruptive, that would be even better.'

Don Henninger and his team are planning another experiment, in a much larger chamber, starting in 2005. But the Chamber will have to have other refinements before it is acceptable for an extended stay on the Moon. Although the inmates were not totally self-sufficient, this is well within the capability of the sort of technology being developed at the Johnson Space Center. Eventually, perhaps, a lunar base will look something like Biosphere 2, whose inhabitants really were self-sufficient. The logistics of setting up an installation like Biosphere 2 on the Moon are daunting, to say the least. Not only would the space colonists need to transport thousands of tonnes of raw materials for building their lunar terrarium, they would have to take water and air. The water could be solidified, and the air liquefied to save space, but their lift-off weight would still be the same. To take all we need to set up a Moon base or Moon city would therefore be a mammoth task, and not one that could be undertaken quickly or cheaply.

It would be much easier, cheaper and quicker if, instead of taking all the materials needed to build a base or a city, we could use what is already there on the Moon. Unlike the Earth – which has rich and varied natural resources – the Moon has no oceans and rivers, no wood, no coal or oil. However, it does have a plentiful supply of lunar rocks and dust. We have seen that there are valuable resources on the Moon that can be used to manufacture spacecraft and electronics, so these materials might also be useful for constructing the buildings for a lunar base. Dave Criswell says that you can 'grab a hold of the lunar materials and change them into virtually anything that you are used to producing here on Earth'. His lunar–solar power idea would depend upon this: it would consist of plots with 'solar cells, wiring and microwave generators and reflective antennas; all of that material – or virtually all of that material – is made on the Moon from the lunar dust'.

The Moon consists largely of basalt, which is also common in the Earth's crust, though not deeper down, in its mantle or its core. This fits in with the theory that the Moon was created after a collision of Earth with a huge interplanetary object: only materials from the Earth's crust, not the mantle or core, would have been thrown into space by this collision. Basalt is basically silicon dioxide combined with various metallic elements.

There are two main types of geological region on the Moon: the highlands and the lowlands. The highlands – the lunar mountains and the crater rims – were formed from materials thrown out from the Moon's crust as a result of the impacts of meteorites. And so they are made from the materials of the Moon's crust: basalt high in aluminium. The lowlands – the maria, the 'seas' – are younger areas, formed by the intrusion of molten rock that originated from far beneath the Moon's surface. The molten rock was probably formed by huge meteorite impacts, generally about 3,000 million years ago. Lowland rocks are also largely made of basalt, but with more iron and magnesium and a little less aluminium; a significant amount of titanium is found, too. There is plenty of scope in these two types of geological formation to make strong metal alloys – from the iron, magnesium, aluminium and titanium. To make iron into steel, lunar smelting works would have to be set up. However, there is a snag: steel-making requires carbon and oxygen in addition to iron. There is very little carbon present on the Moon, and huge quantities would have to be taken there. Clearly there is no oxygen gas in the atmosphere – the Moon has no atmosphere. However, oxygen could be liberated simply by roasting the silicate minerals contained in the lunar basalt. This oxygen could provide life support for human inhabitants, too, or oxidizers for rocket fuel. The energy to roast the silicate would come from the reliable supply of solar radiation. The silicate minerals would also be useful for making glass, fibreglass and ceramics. Some of the necessary materials that are not present on the Moon – in particular, carbon – are found on asteroids, and they could be mined from the asteroids and brought to the Moon.

Tireless workers

Naturally, if a base is to be constructed on the Moon, a minutely detailed survey will be needed before any building work can begin. First, though, the construction workers would have to be sure that they were in a location where they could find the right materials. Once all this was done, the lunar minerals would have to be processed, to make the glass and metal alloys from which the buildings could be made. It would be difficult for people to do this, as they would have to live on the Moon during these long early stages. But robots could be sent up to carry out these tasks.

Professor Kumar Ramohalli has clear ideas of the way this would work. He is based at one of the Space Engineering Research Centers set up by NASA at several American universities. At the University of Arizona, Ramohalli has designed several robots for exploring the Moon or Mars, and utilizing the resources there. These include LORPEX (Locally-refuelled Planetary Explorer) and BIROD (the Biomorphic Robot with Distributed Power). LORPEX would manufacture its own fuel from the lunar dust, using solar power; then it would use that fuel in its rocket engines, to make powered 'hops' so that it could explore other regions. BIROD uses some recent developments of robotics, including artificial muscles, which allow it to move around more like a cat than a wheeled robot. Its muscles are powered individually, rather than centrally – just like real muscles, which derive their energy from chemical reactions in the cells of which they are made. It is small enough to fit inside a spacecraft, and could be easily deployed on to the surface of the Moon or Mars.

Robotics is a rapidly advancing field of endeavour. Ramohalli says that 'the Moon will be teeming with intelligent robots during the next thirty years'. Ramohalli – and other researchers – have already made building blocks by baking clay made from lunar-simulant dust. The resulting material, from which the blocks are made, is called 'lunacrete' – the first Moon bases could be fashioned from these lunacrete blocks. Ramohalli has worked out the processes by which a self-contained robot spacecraft could make these bricks. His 'Common Lunar Lander', proposed in 1992, lands softly on the Moon, and its robot arm picks up soil, puts it into a solar oven, and uses the intense heating effect of concentrated sunlight to make bricks and tiles. He believes that a single robot could build twelve huts in six months.

Another robot being developed to play a role in lunar exploration is Nomad, developed jointly by NASA's Intelligent Mechanisms Group and a team from Carnegie Mellon University, Pittsburgh. In a trial conducted in the Atacama Desert, in Chile – over terrain similar to that found on the Moon – this four-wheeled robot travelled 215 kilometres searching out particular types of rock, and operating largely autonomously. The robot had video cameras that gave 360-degree vision, and a sophisticated communications capability, which meant that it could be controlled remotely by scientists more than a thousand kilometres away. It was also intelligent enough to be able to explore and make decisions for itself. Red Whittaker, one of the team who developed Nomad, says that 'Nomad has capabilities to see, to safeguard, to communicate ... it is a continuing persistent evolution of the technologies and the operations and the ideas that are essential to carry us back to the Moon.'

Nomad is equipped to carry out geological experiments, and to work out the nature of and distances to nearby mountains or boulders. For some portions of its trek through the Atacama Desert, the images from the panoramic cameras were displayed on an 11-metre-wide curved screen at the Carnegie Science Center in Pittsburgh. Visitors to the centre could control the vehicle as if they were in it – a technique called telepresence. In the future, telepresence systems may use virtual-reality systems, allowing a user on Earth the chance to 'look around' the Moon and pick things up using a virtual-reality glove. There would be a serious time delay, however, as signals between the Earth and the remote robot would take nearly two seconds each way. The whole robot has a mass of 550 kilograms, which means it could be taken to the Moon relatively easily. One of the clever design features of the Nomad robot is the fact that its wheels retract. The robot is 2.4 metres wide when it is roving – this wide wheelbase gives it stability. So that it can be stowed in the hold of a spacecraft, the four wheels retract under the robot's body, making the robot 60 centimetres narrower. Another member of the Nomad team is Dimi Apostolopoulos, who explains a rather strange attachment to the robot: 'We all have a very intimate relationship with this robot. We are the people who conceived it, designed it, built it – when Nomad is on the Moon, it will be as if one part of ourselves is actually operating up there.'

The idea of robots building lunar bases for human habitation sounds like a scene from a science-fiction film. But if the incentives are there, the technology is sure to follow. Kumar Ramohalli may be right: the Moon really could be teeming with intelligent, industrious robots in the near future. And so, with robots surveying the Moon's surface and then building the initial primitive shelters for humans, the first permanent lunar base could be established within a few years. By living in units like the Chamber, humans could survive on the barren Moon for extended periods. But they would still need to take essentials like water and oxygen – or would they?

Where the Sun don't shine

It has long been assumed that the Moon is totally dry: because it has no atmosphere, and because every part of its surface receives direct sunlight, any ice there would quickly evaporate into space. However, in 1961 three lunar scientists at the California Institute of Technology in Pasadena proposed that water ice might exist at the north and south poles of the Moon. And over the next three decades other scientists developed this idea, suggesting that ice could be deposited in craters by comets that crash on to the Moon. Comets are made of ice and dust – they are sometimes referred to as 'dirty snowballs'. It was noted that there are certain craters at the Moon's poles whose interiors never receive any sunlight. In these shadowed areas, the temperature would never rise above minus 170 degrees C. In 1992 and 1993, in an effort to investigate the possibility of ice on the Moon, radio waves transmitted from Earth bounced off the lunar poles and were detected by Earth-based radio telescopes. The results seemed to suggest that water really might be there.

In 1994 NASA and the US Department of Defense launched a space probe, Clementine, which was supposed to travel – via the Moon – to an asteroid. It never made it to the asteroid because of a fault in its on-board computer software, but it made an astonishing discovery as it flew past the Moon. It shone radio waves on to the Moon's south pole, and the reflected signals were received and analysed on Earth. The results were extremely encouraging – they seemed to indicate the presence of thousands of tonnes of water – until they were shown to be inconclusive. More information was needed to determine whether or not there is water on the Moon. If there is, then lunar survival will be a great deal easier.

In 1998 another tireless worker – this time an orbiting probe called Lunar Prospector – set out to further investigate Clementine's discovery. Lunar Prospector carried a neutron spectrometer, which would measure the energy of neutrons produced at the Moon's surface – these neutrons are expelled when cosmic rays collide with the Moon. The energy of the neutrons depends upon what substances are present on the Moon. So Lunar Prospector could not only search for signs of water, but could also perform a general survey of the distribution of resources on the Moon as a whole. Alan Binder, based at the Lunar Resources Institute in Tucson, Arizona, was the principal investigator on the project. He says, 'We're not just doing science: we're trying to understand the distribution of resources on the Moon which we can then use to build a lunar base, to build a lunar colony.'

Lunar Prospector made a total of 6,800 orbits of the Moon, over a period of eighteen months. And its neutron spectrometer did find fairly conclusive evidence that there were hydrogen atoms in huge quantities in the craters at the Moon's poles. The hydrogen atoms are almost certainly connected to oxygen atoms, in water molecules (H_2O). So Lunar Prospector has confirmed Clementine's results – there really is water on the Moon. The total mass of ice in these craters is not known, but is probably somewhere between ten million and 500 million tonnes. This would be enough to support a colony of about a thousand people for about a hundred years – or much longer with efficient recycling. In a dramatic attempt finally to confirm the existence of water, the Lunar Prospector was deliberately crashed into a crater at the south pole – only when its mission had already overrun and its batteries were running low. The hope was that a plume of debris would be produced by the collision: the 160-kilogram craft was travelling at 6,120 kilometres per hour when it hit. Earth-based telescopes, as well as the Hubble Space Telescope, were watching as the craft crashed into the Moon on 31 July 1999. The telescopes were looking for sunlight reflected off the debris. By splitting that light into a spectrum, astronomers would be able to tell whether water was present.

Despite the earlier positive results, these observations did not find any water in the debris. It is possible that the spacecraft simply missed its target, and hit part of the Moon's surface outside the crater, or that the debris was not carried high enough to be detected – or, of course, that there is no water after all. It may be that the hydrogen detected by Lunar Prospector was an accumulation of hydrogen ions (electron-less hydrogen atoms) from the solar wind, known to be streaming out into the Solar System from the Sun, as described in 'Sun Storm'.

Would-be lunar explorers are very excited about the possible presence of water on the Moon. Not only will water be useful for drinking, washing and agriculture: using solar power, it is possible to separate the hydrogen and oxygen that makes up water, to make fuel for rockets in which to travel back to Earth or farther out into space. Alan Binder believes, 'Everybody likes to imagine themselves being on another planet, exploring the Solar System. This is now at the point where it can happen – it will happen; now that Prospector has found water, we can have a golf course and a swimming pool.'

Many organizations are busy putting forward their ideas of what we can do with the Moon. The first aim is to find out yet more information about it using uncrewed space probes. Several lunar probes have been planned for the near future, including Icebreaker, being developed jointly by a private company, LunaCorp, and the Robotics Institute at Carnegie Mellon University. If their probe is launched, this wheeled robotic vehicle will survey areas around the south pole, looking for water, between June and December 2003. Icebreaker will have a drill and ground-penetrating radar, so that it can look for water in the regolith underneath the surface. The Japanese Institute of Space and Astronautical Science has put together a thirty-year project for a lunar base. The first stage of this project is planned for 2002, when a probe called Lunar-A will be launched to investigate what is underneath the regolith, using seismic surveys. Another proposed Japanese project is LOOM (Lunar Orbital Observatory Mission) a 2,000-tonne space probe that would map out the lunar surface from an altitude of 100 kilometres.

At least two European missions to the Moon are currently being planned. One team, based at the Technical University of Munich, has proposed a low-budget lunar orbiter, called LunarSat, which would fly over the Moon's poles investigating in greater detail the distribution of ice. Other scientific plans that have been put forward include building a huge observatory on the Moon. A telescope that does not look through the atmosphere of Earth – such as the Hubble Space Telescope currently in orbit around Earth – has significant advantages over Earth-based telescopes. Observatories with telescopes based on the Moon rather than in orbit would be easier to update and repair. The European Space Agency (ESA) is among the organizations that plan to set up lunar bases, as part of a programme called the Lunar European Demonstration Approach.

Everyone's going to the Moon

Moon fever is taking hold, with private companies becoming increasingly active in the drive for development. There is LunaCorp, based in Arlington, Virginia; they are running several programmes jointly with the Robotics Institute at the Carnegie Mellon University, including Icebreaker, mentioned above. In another of these joint projects, LunaCorp intends to send two interactive robot rovers to the Moon. People on Earth will be able to sit on 'motion platforms' and experience the same knocks and bumps that the rover is experiencing on the Moon. Some users will be able to control the robot probe, and Internet access will enable anyone with an Internet connection to obtain live pictures from the Moon. The probes will have Internet-ready computer servers built into their wheels, and will communicate via radio like any other probe. This and other aspects of the commercialization of the Moon were explored at the first Commercial Lunar Base Development Symposium, held in July 1999 in League City, Texas. It was organized by the Space Frontier Foundation and funded by the Foundation for the International Non-governmental Development of Space. Things really are happening.

Greg Bennett of Bigelow Aerospace and the Lunar Resources Company gave a talk at the symposium, entitled 'Lunar Cruiseships and Hotels'. The main thrust behind the Lunar Resources Company is the Artemis Project, which aims to establish a permanent, inhabited and self-sufficient base on the Moon within ten years. Once there, its lunar pioneers will explore the Moon to find the best locations to set up mining operations and a permanent lunar colony. The money to do this will come from the initial stages of space tourism: taking people for thrilling rides into space or perhaps around the Moon. In the foreseeable future, say the Lunar Resources Company, the Artemis Project will include travel to and from the Moon in huge spaceliners and a stay at a luxury hotel on the Moon itself. In his work for Bigelow Aerospace, Bennett is involved in visualizing how hotels might work in space. The owner of the company is tycoon Robert Bigelow, who has pledged to support the building of a 100-passenger tourist cruise ship that would orbit the

Moon; he has promised $500 million to the project, before it starts making him money. Bennett can also foresee tours around the original Apollo landing sites, where visitors walk around in transparent plastic-covered walkways.

Patrick Collins, too, feels that the future of the Moon will depend upon space tourism. He believes that the first people to go there are going to be 'adventurers – people for whom it's really their life's ambition; and then it will be the rich who go, once it's really safe; and then if you go there in a hundred years, it will be like Vegas'. Despite the fact that it is in the middle of the desert, Las Vegas is an incredibly rich city, and is growing richer all the time. The money comes from tourism. Collins is certain that the same demand will exist for space-based tourism: 'Frankly, the world is getting smaller and smaller now; there is nowhere you can go that's new; there's nowhere you can go where you are out of reach of a mobile phone. So space is the next frontier for tourism.'

Most people would think that spending hundreds of thousands of dollars to go to the Moon just to avoid your mobile phone is a little excessive, when simply turning it off would be cheaper and easier. However, Collins is probably right about the future of the Moon being dominated by tourism. The Moon draws people towards it for many reasons – space scientists have a million questions; adventure-seekers would find one of the ultimate thrills; artists dream of painting there; and of course people want to mine the Moon's resources. And if people like Patrick Collins and Greg Bennett are right, then tourism is the only thing that can make these things possible. Space tourism, however, will be available only to those people with huge amounts of spare money, who will be very much in the minority. Long-haul air travel suffers from the same inequalities today: the majority of the world's people will never be able to afford to take an intercontinental holiday, and yet some people can travel anywhere in the world and be pampered in luxury hotels. Some people would argue that long-haul air travel would not have been developed if it were not for capitalism in general and tourism in particular. But then, just what value long-haul travel and expensive hotels have to those who will never be able to afford to experience them is another question.

Patrick Collins and his fellow 'space profiteers' are serious. If their visions come true, then perhaps thousands of people will be visiting lunar hotels each year by around 2040. What might a visit to a hotel on the Moon be like?

Even in 2040, your spacecraft has to produce great thrust to accelerate to the speeds necessary to lift you into orbit. So its huge rocket engines start up and lift you into space – as the spacecraft accelerates, your body is pulled up with it. And so you feel yourself being pressed harder against your seat. The downward force is 'g-force': at '3g', you push against the seat three times as hard as your weight would normally do if you were not accelerating upwards. You experience the same thing when a lift begins to ascend inside a building – your legs bend, and you press harder than normal against the floor of the lift. If your spacecraft has windows, you can see the sky darkening outside you as the atmosphere thins. As you move into orbit, the g-force decreases, and eventually you experience less than the 1g that you are used to. In fact, once you are in orbit, your spacecraft and you are both free-falling, in zero-g. You are falling down towards the ground, but you never move any closer to it. This is because you are moving around the Earth in orbit – always at a right angle to your direction of fall – at just the right speed. Once in orbit, you are moving at incredible speeds, but as you gaze out at the Earth, it looks as if you are just drifting slowly. This is simply because you are so far above the Earth's surface, which provides your only visual reference. It does not feel as if you are moving fast either, because at a constant speed your body feels no g-force. In fact, you are now experiencing weightlessness – although you will probably remain strapped into your chair for some time yet.

Your spacecraft acts only as a shuttle to a much larger space cruiser that ferries people between the Earth and the Moon. Your craft docks with the cruiser, and takes this opportunity to refuel. It makes sense to take fuel down to Earth ready for the next flight up into orbit. The fuel taken on board by the shuttle craft has been transported in the cruiser from the Moon, on the return leg of its last trip. It may have been produced by the action of solar power on lunar soil or – if enough ice has been found on the Moon – by splitting water into hydrogen and oxygen. You and your fellow passengers transfer – in zero-g – from the small craft to the large space cruiser, through an extremely well-sealed airlock passage. Your luggage is transferred to the larger craft, too – though the amount you are permitted to carry is very limited. After you board, there is a detailed safety briefing and then you are strapped into your seat again, ready for the push towards the Moon. A carefully calculated rocket thrust sends your space cruiser out of Earth orbit, and into a lunar trajectory.

The space cruiser was built, in orbit, several years previously. In the vacuum of space, it does not need a streamlined shape – as a seagoing cruiser does – if it is to move quickly towards its destination, and so this ship can have any shape. However, most of the modules from which the space cruiser is made are cylinders – because that shape is strong enough to withstand the incredible pressure difference between the inside of the cruiser and the vacuum outside. The cylindrical modules are quite spacious, and there are huge windows through which you can enjoy the view, as Earth grows smaller and you head for the Moon. For the next four days, you enjoy luxury in weightlessness. You can communicate with people on Earth, as the ship is connected to whatever the Internet has become by 2040. You can watch television, play games, or just enjoy the weightless conditions as you travel; there may be a gymnasium, too. Perhaps part of the space cruiser will be set into rotation, so that normal gravity is resumed in this area of the ship. And so after four days' travel, your space cruiser manoeuvres into a lunar orbit and you transfer to another shuttle craft, this time destined for the Moon's surface.

From your vantage point above the lunar surface, you can just make out a source of light near to the Moon's north pole. It is a huge dome made of glass and aluminium, about one kilometre in diameter. Around it, there are fields full of crops inside other huge glass domes. In their experiments in the 1990s, the team at the Kennedy Space Center worked out that about thirty square metres of land would supply each person with food – if genetically modified, high-yield plants are used. This is about half the area of a squash court. Next to the agricultural areas is a collection of scientific laboratories, including several astronomical observatories. About a hundred kilometres from the settlement – beyond the lunar mountains that lie next to the huge dome – are extensive arrays of solar cells and a mining operation. The shuttle craft's rockets fire to slow your approach to the settlement and the craft hovers before landing at the end of a covered tunnel, which extends about six kilometres from the main dome. You transfer from the shuttle craft, through another airlock, into the tunnel. Much of the length of the tunnel is transparent, so you can see the lunar surface, the black starry sky and perhaps the Earth, as you are transported towards the main dome.

Just like a large version of the Biosphere 2 project in the 1990s, the dome is sealed and acts as a self-contained ecosystem. All the resources inside are carefully monitored, although the recycling regime is not as strict as in Biosphere 2 – fresh oxygen can be obtained from the lunar soil, for example, and other essential substances can be mined from asteroids thousands of tonnes at a time and brought to the Moon. Besides, a strict recycling regime may prove uncomfortable for the hotel visitors, who would not want to pay large sums of money to live meagrely during their time on the Moon. The temperature inside the dome is carefully controlled, to avoid those dramatic temperature swings experienced by the exposed surface outside the dome. The glass of the dome is designed to filter out much of the ultraviolet in the sunlight, just as the atmosphere does on Earth, though you can still sunbathe. One of the buildings inside the dome is the hotel; the lighting inside it varies over a twenty-four-hour period, to simulate the cycle of day and night on Earth.

When you arrive in the dome, you are taken to the reception area of the luxury hotel, which is similar to the rich palatial hotels found on Earth. Water falls slowly to the ground in the low gravity of the Moon, so in the hotel foyer there are graceful water features unique to the Moon. There are also trees and shrubs growing all around you – the foyer is grand and welcoming. One of the best things about being on the Moon is the low gravity. For example, people are able to leap as high as their own height on the Moon. There is a range of excursions that you can take to see the sights of the lunar surface – including a tour of the Apollo landing sites, where the lunar pioneers stood some seventy years ago. And of course, there is a gift shop, and you will find a wealth of 'photo opportunities'.

Back to the present

At the moment, this chain of events is a fantasy. However, there is nothing in the story that, in principle, cannot be achieved using technology that exists today or will be developed in the very near future. And there are already serious, costed proposals for a lunar hotel complex in a huge glass dome. In 1998 it was reported that the Hilton Hotels Group had commissioned architect Peter Inston to look into the idea of building a hotel on the Moon. In the same year, Japanese construction company Shimuzu also announced plans to build a lunar hotel. Even if these particular projects are never undertaken, more will follow. Patrick Collins has clear visions about what a lunar hotel might be like: he says that the 'ceilings of the rooms are going to have to be higher, because in the one-sixth gravity on the Moon, when you walk you bounce up and down'. And Greg Bennett supposes that the effects of low gravity could 'change the way you design staircases'. He also believes that in a lunar hotel, you could be 'standing on floors made of the actual rock of the Moon'. That is an exciting prospect – if you can afford it.

The Moon is big business, even though no plans have actually made it off the ground yet. There is already a lunar estate agent, called Lunar Embassy, that sells small plots of land and even city-sized regions of the Moon's surface. Lunar Embassy claims that its operation is legal, and has already formally founded a city called Lunafornia, too. There are plenty of private companies and organizations keen to kick-start lunar development – what about public-sector organizations, or some kind of international regulatory body to govern the future of the Moon? The closest thing is the International Lunar Exploration Working Group, formed in 1995. This is made up of representatives of most of the major space agencies, whose mission is to develop a strategy for lunar exploration, with international co-operation as a priority. Their role may be crucial in helping to define the shape of future developments.

The Moon will always be important scientifically, being our nearest celestial neighbour and consisting of important ancient minerals that may tell us a great deal about the formation of the Solar System. But the Moon will be important to space scientists in another way, too. Because of the low gravity and lack of atmosphere, it is easier to plan, equip and launch spacecraft bound for Mars or Jupiter from the Moon, and it would be much cheaper and easier than launching it from the Earth. So a lunar base may play a central role in our quest to explore the other planets. As we have seen, the interest in the Moon is certainly not restricted to the scientific community. Lunar miners want to dig up huge areas of the Moon, and will need to cover vast areas of the lunar surface with solar cells. And lunar hotels will be set up, if they are guaranteed to make money. Once enough people are working there, bases and hotels may become integral parts of a new civilization.

Alan Binder has a clear vision of the Moon's future: 'Thirty years from now, I can imagine tens of thousands, hundreds of thousands of people living on the Moon ... you could have them stay for ever, and have children, and have their children have children.' For Steve Bennett of the Starchaser Foundation – one of the entrants for the X Prize discussed in the previous chapter – the Moon is 'important for the long-term continued human exploration of the Solar System, because it's filled with all the important things that humans need to live, work and prosper in space'. If a colony grows on the Moon, through lunar exploration and exploitation, then the next logical step would be to set up bases, mines, hotels and colonies on Mars.

The red planet has some features that would make it easier to colonize than the Moon. For example, it has an atmosphere – though not one in which human life could survive. The factors that make Mars uninhabitable at present are its surface temperature and the nature of the atmosphere. These two factors are related: Mars is colder than the Earth not only because it is farther from the Sun than the Earth is (though that is important too). It is also because it has a more tenuous atmosphere than Earth; the atmosphere on our planet prevents energy from leaving the planet as infrared. This is the greenhouse effect, and makes a significant difference to the Earth's surface temperature.

Some long-term plans suggest terraforming the planet: changing the Martian atmosphere to make it like our own. The first step would be to warm the polar ice caps, which are a mixture of water ice and carbon dioxide ice – one method of doing this would be to mount huge mirrors in space that would focus sunlight on to the Martian poles. The water and carbon dioxide would evaporate and form an atmosphere. Through the greenhouse effect, the presence of an atmosphere high in carbon dioxide would warm the planet; then plants could be brought in, changing most of the carbon dioxide into oxygen as they grow, making the atmosphere breathable. This really is possible – in the future, there may be millions of people living on Mars, not confined to sealed domes but living normally in the open air. There is little chance of you setting up home on Mars: this whole process would probably take centuries.

Today, the long-term future of human beings in space is uncertain. Jim Benson of SpaceDev foresees a commercial future in space, filled with many possibilities: 'I think we can accomplish anything that our imaginations allow us – and that the technology is available for – and it's so hard to predict what can be possible; but why not have orbiting businesses and hotels leading to settlements on the Moon, and scientific outposts and even hotels on the Moon and Mars eventually? All of this is possible, and will happen; the question is when?'

With so many problems still to solve on our own planet, the question will also always have to be: 'Why?'

SPACE – THE
FINAL JUNKYARD
... searching for solutions...

If you are interested in the future of space exploration, or
might want to take a trip into space yourself, there is
something you need to know. Space has become a junkyard, full of debris from thousands of space missions. This
debris is not merely an unpleasant littering of our space
environment: it is a threat to the safety of future space missions. Space junk can damage spacecraft or break through
an astronaut's protective spacesuit. It can even be a danger
to people back here on Earth.

Floating free

A story from the early years of space exploration will help to
illustrate the important features of space debris, some
unusual pieces of which were created during the mission. On
the morning of 3 June 1965 astronauts Ed White and James
McDivitt lay back in their seats in a capsule atop a Titan II
rocket. The rocket fired and accelerated upwards. As the
rocket climbed, its first stage fell away; the second stage continued into space and pushed the crewed capsule into orbit
around the Earth. The average altitude of the orbit was 223
kilometres. In this orbit, capsule and crew were travelling
at speeds of more than seven kilometres per second. Soon
after the spacecraft achieved orbit, the second stage of the
rocket was released. It was also travelling at just over seven

kilometres per second and so, after it was released, it continued to orbit with the capsule.

After a few minutes, the crew estimated that the second stage was about a hundred metres in front of their capsule. One of the aims of this mission, called Gemini IV, was to attempt a rendezvous with the second stage of the rocket. This would be important for the impending Apollo missions: astronauts who descended to the Moon's surface in the lunar module would have to ascend again to meet the Moon-orbiting command and service modules. During Gemini IV's attempted rendezvous, McDivitt fired the capsule's rockets, intending to push the craft in the direction of the second stage. Frustratingly, he found that instead of moving closer to his target, he moved above it and behind it. This is because the increase in speed sent the capsule into a higher orbit, which takes longer to complete than a lower one. So firing the rockets to take the capsule towards its goal actually allowed the free-floating second stage to move further ahead relative to the capsule. To achieve rendezvous with an object ahead of it, a spacecraft has to slow down – this drops it to a lower orbit, in which it can catch up whatever it aims to meet. When it has caught up, it moves back into the higher orbit and the rendezvous is complete. All this looks like a graceful, slow-motion manoeuvre. However, it is important to bear in mind that the two orbiting objects are actually travelling at incredible speeds – this is more like a high-speed chase.

The second aim of the Gemini IV mission was just as important. After three revolutions in orbit – each taking about ninety minutes – White was ready for what was probably the most exciting moment of his life: time outside the capsule, attached to it only by a 7.6-metre tether. As well as providing a guarantee that White would not drift away hopelessly into space, the tether included an oxygen hose and other vital links, making it a lifeline in more ways than one. White used a 'personal propulsion unit', or 'zip gun', to move about in space. Pulling the gun's trigger released oxygen under pressure, pushing the astronaut in the opposite direction. The only other way he could have moved away from the spacecraft was by pushing off it – this could have disturbed the orbit.

There is an important point here: when White let go of the capsule to float free, he did not get left behind, even though the capsule was travelling more than seven kilometres per second. The reason for this is that White, too, was travelling at that speed. If you jump from an aeroplane, in the atmosphere, you do get left behind. The difference is due to air resistance. As someone steps out of an aeroplane, they are travelling at the same speed as the aeroplane at first, but moving through the air slows them down, so the aeroplane speeds off without them. The reason the aeroplane carries on moving at the same speed is that its engines provide thrust that pushes it forward despite the air resistance. In orbit several hundred kilometres above the ground, there is virtually no air; there was nothing to slow down White or the spacecraft – and so they orbited together at high speed. In television footage of astronauts on spacewalks it looks as if the spacecraft and the astronauts are drifting slowly, floating almost motionless above Earth. But like the discarded second stage of the Titan rocket, and like White and his capsule, they are actually moving at around seven kilometres per second. This is the main reason why space junk is dangerous – any debris that is left in orbit by a space mission will be travelling at this kind of speed. At seven kilometres per second, a small nut or bolt becomes a lethal projectile that can kill a person or rupture a spacecraft.

During his spacewalk, Ed White was outside for fifteen minutes and forty seconds. While he was above the Pacific Ocean, tethered to the spacecraft, he discarded his outer glove and his helmet visor. These unusual pieces of space junk drifted away from the spacecraft, and probably remained in orbit for several months. Since this first American spacewalk – and the first Russian one three months previously – about a hundred more spacewalks have been taken. During most of them, items have been discarded or have slipped from an astronaut's grasp. Spanners and cameras have been lost, for example. As well as creating their own debris, the crew of the Gemini IV mission may have narrowly missed a deadly collision with another piece of space junk. White and McDivitt reported seeing an unidentified flying object hurtling towards them. Although McDivitt appeared on television many times claiming that the object was a flying saucer, it was probably a piece of debris from another mission.

Although they were travelling at such incredible speeds, White's glove and visor posed no threat to the capsule, as their speed relative to it was very low indeed. In fact, all satellites (apart from those in a polar orbit, described below) orbit in the same direction: from west to east. This is chosen because it is the same direction the Earth itself turns – the Earth's rotation means that a spacecraft is already travelling west–east at 400 metres per second before it even lifts off. Because everything is moving around the Earth in the same direction, head-on collisions between objects in orbit thankfully do not occur. Also, two objects at the same height will be travelling at about the same speed, reducing the risk of collision. Imagine two athletes running in the same lane of a running track; if they are travelling at the same speed, they will not collide, and if there are no other athletes coming in the opposite direction, no one else will collide with them. However, most satellites do not orbit parallel to the equator, so objects in orbit may cross each other's paths, and can then collide. Similarly, if the two athletes cross into each other's lanes, they could collide, but not with any great force.

A similar situation occurs because most orbits are not perfectly circular. Gemini IV's orbit, for example, had an apogee (maximum altitude) of 282 kilometres and a perigee (minimum altitude) of 163 kilometres. Again, there is a chance that objects in different orbits will collide. Most dangerous of all, some satellites orbit over the poles and so cross many other orbits at 90 degrees. Imagine two sprinting athletes in running tracks set at 90 degrees to each other – in this case, the collision could be dangerous. At the speeds that objects orbit in space, it could be catastrophic.

Space jam

While collisions between objects in space are rare, they do happen – and their frequency has increased as more and more objects have been left in orbit. Since the end of the 1950s, nearly 5,000 satellites have been launched into orbit. Only about 350 of them are still working, and many others have returned to Earth. Don Kessler, an expert on orbital debris, says that there are two reasons why debris has accumulated to such a degree: 'One is that people were not aware of what was going on, and second, there's no economic incentive not to leave something in space after you're through with it.' Nick Johnson of NASA's Orbital Debris Research Project estimates that 'the total mass in Earth orbit … is approximately 5,000 metric tonnes, and is growing at a rate of almost 200 metric tonnes every single year.'

There are more than 8,000 objects larger than ten centimetres across, and countless smaller ones. Among this motley collection are satellites that have either malfunctioned or simply come to the end of their working lives. Then there are rocket bodies – engines and fuel tanks. For objects being placed into a low Earth orbit, usually only one part of the rocket makes it into space – the final stage. However, satellites that are destined to circle the Earth at higher altitudes, such as those that end up in geostationary orbits, arrive in space with two or three. Another source of debris is components of spacecraft such as nuts and bolts and, as mentioned earlier, items dropped by astronauts during spacewalks. The exhaust gases of rockets fired in space can produce huge numbers of tiny particles. Finally, there are fragments of spacecraft produced by break-ups. Many break-ups are the result of explosions, which usually happen because residual rocket fuel mixes with liquid oxygen or because pressure becomes too high inside a satellite's battery.

Collisions involving space debris also lead to break-ups that produce more debris. One of the first satellite break-ups that is attributable to a collision with space junk occurred in 1981, when a Russian navigation satellite called Kosmos 1275 blew up in space. Orbital-debris scientist Darren McKnight explains that 'the interesting thing about Kosmos 1275 is that there was really no energy source on board that should have caused the satellite to break up into 300 pieces of debris ... Mainly, people have come to the conclusion that they think it may have broken up due to a collision with another piece of orbital debris.'

Collisions themselves can sometimes cause explosions in fuel-laden satellites, as well as cause break-up directly. As a result of all these explosions and collisions, there are millions of fragments smaller than tennis balls. Some of the spacecraft break-ups that have created the smaller fragments of space junk were actually planned. From the late 1960s until the early 1980s, the USA and the then Soviet Union conducted tests of 'star wars' weapons that would approach orbiting enemy satellites and fire high-speed pellets at them. The tests created thousands of fragments. Yet another type of debris is coolant leaking from damaged satellites. At present, there are at least thirty Russian satellites leaking liquid sodium and potassium that was used to cool the nuclear reactor on board. At altitudes of between 800 and 100 kilometres, there are clouds of these droplets.

Below an altitude of about 3,200 kilometres, nothing escapes collision with space debris. Evidence of the dangers created by space junk can be seen on the solar panels and other surfaces of satellites recovered from orbit, which show pockmarks of various sizes caused by bombardment by thousands of small particles. By analysing the pockmarks on spacecraft retrieved from various altitudes, researchers have been able to estimate the distribution of space junk in different orbits. The highest concentration of debris is found in low Earth orbits, below about 600 kilometres.

Every piece of space junk is travelling at the phenomenal speed necessary to keep it in orbit. These speeds – several kilometres per second – are called hyper-velocities. When moving at hyper-velocities, an object just one millimetre across can inflict the same damage as a high-speed rifle bullet at close range. A 1-centimetre hyper-velocity object has the same energy of impact as a car travelling at 100 kilometres per hour, or a 1-tonne metal safe dropped from about thirty metres. A 10-centimetre object has about the same energy as the explosion of several sticks of dynamite. Even the droplets of coolant can be dangerous; Don Kessler says that for objects that hit at hyper-velocities, 'whether they're liquid or solid doesn't really make any difference'. Space has become a cluttered and dangerous place to be.

The collision between one of the larger particles and a spacecraft can cause the craft to tumble out of control or stop working altogether. Experiments conducted at the University of Texas show what can happen in one of the worst cases – when a projectile hits a spacecraft at a hyper-velocity. These experiments involve a 'light gas gun', which uses extremely high-pressure hydrogen or helium gas to accelerate small pellets to speeds of several kilometres per second. In one such test, a satellite model made of thick aluminium is placed inside a thick-walled chamber and a plastic pellet is fired at it, at a speed of 6 kilometres per second. Soon after the collision, hyper-velocity scientist Dr Harry Fair enters the chamber to investigate the damage. The model satellite is totally destroyed: the thick aluminium sheets that made the body of the satellite are buckled and are still hot after the collision, and the electronic circuit boards are charred and broken. Fair says that the impact spelled catastrophe for the satellite, and that 'it would be very difficult to conceive of anything surviving after that kind of impact'.

Similar experiments are carried out at NASA's hyper-velocity laboratory at the White Sands Test Facility in New Mexico. The researchers there use light gas guns to fire pellets up to 2.5 centimetres in diameter at 7.5 kilometres per second. The pellets slam into test pieces of various materials, including pieces of multi-layered spacesuit and new types of spacecraft shielding being developed at the laboratory. High-speed X-ray photography highlights just what can happen as a result of hyper-velocity impacts. Frame-by-frame, researchers watch pellets vaporize as they break through a thick aluminium plate, producing a spray of tiny particles. If spacecraft were made of a single layer, then a collision like this would rupture the hull. The particles in the test facility go on to produce pits and cracks in the wall behind the aluminium plate. Multi-layered shielding for spacecraft is being developed at the White Sands facility – each layer absorbs some of the energy of any particle that hits it, and the particles are vaporized before reaching the spacecraft's hull. This research is important for understanding not only impacts involving space debris, but also those involving small meteoroids – natural high-speed pieces of interplanetary rock and dust that can present as much threat as the debris of artificial satellites. When a micrometeoroid vaporizes on impact with a satellite, for example, it produces a plasma – a high-temperature gas of charged particles – that can damage a satellite's electronic circuits.

Whether meteoroid or space debris, there have been several known collisions with spacecraft since the demise of Kosmos 1275 in 1981. In 1996, for example, a French military satellite called Cerise was put out of action when an object the size of a briefcase collided with its stabilizing mast, sending the end of the mast hurtling away. Since the job of the mast was to keep the satellite pointing in the right direction, the satellite began tumbling hopelessly out of control. Operators on the ground have now managed to regain control of the satellite, although its operation has been seriously impaired. Many other satellites have suddenly stopped working for no apparent reason – most of these were probably the result of collisions with space junk. And there have been plenty of known near-misses. In 1999, for example, an American military satellite came within 500 metres of the Russian Mir space station, which had come to the end of its life and been abandoned a few weeks earlier (although it has now been promised a new lease of life by an international investor). During Mir's long and illustrious career in space, the crew had to move into their escape capsule eight times because of potential collisions with orbital debris.

A smash of paint

In one of the early Space Shuttle missions, in 1983, a tiny fleck of paint from another spacecraft hit one of the orbiter's windows, creating a 4-millimetre crater in the toughened glass. This occurred before the space community was totally aware of the dangers of space junk. The Space Shuttle orbiter vehicle is peppered with small collisions during every mission. The windows – costing $40,000 each – must be replaced regularly because of this kind of damage. There is another type of space junk that Space Shuttle crews had to be aware of when they were docking with Mir. Pam McGraw at the Space Shuttle Mission Control explains: 'The Russians tend to throw their garbage out the back door, so when we're approaching Mir, we're flying up through some of that.' In this case, the garbage includes human waste from the toilet facilities of the Russian space station. The waste is contained in capsules, each one an aluminium cylinder about the size of a rucksack. The capsules are deliberately discharged – propelled towards the Earth, so that they burn up re-entering the atmosphere.

When the Shuttle was being designed, in the 1970s, the risk from space debris was not well known. The engineers on the Space Shuttle programme have now installed extra protection against small items of debris. However, larger objects could cause much more serious damage. If you are aboard the orbiter, there is no way you can dodge a chunk of debris using just your senses: neither your senses nor the orbiter will be able to react quickly enough. One of the

Space Shuttle pilots, Scott Horowitz, says, 'You're probably never going to see this particle coming towards you because the closure speed can be in the order of fifty or sixty thousand kilometres per hour – you're just moving so fast.' Ground-based mission control can predict collisions with the larger objects. Horowitz explains: 'We have a whole group of engineers that do nothing but analyse the risk of orbital debris … and change the alignment of the Shuttle to protect the critical systems from orbital debris hits.' The engineers are able to warn the Shuttle crew thirty-six hours in advance about any chunk of debris that will come within ten kilometres of the orbiter, and adjustments can then be made to the orbit. To reduce the dangers from smaller particles, the orbiter tends to fly with its least vulnerable part – the tail – towards the most likely direction from which space junk will hit.

The odds on a satellite colliding with a piece of space debris cannot easily be calculated – particularly for small debris – but estimates have been made. The probability of collision increases with the size of the 'target' and the length of time it is exposed to the junk-filled environment. Some orbits present a greater risk, too. A person is small compared to a spacecraft, and spends only a few hours at a time in a vulnerable position, so the chances of an astronaut being hit are very small. Various agencies have produced estimates of the probability of collision with space junk. According to the Colorado Center for Astrodynamics Research, the probability that a piece of debris larger than one centimetre in diameter will hit a medium-sized spacecraft during a ten-year period is somewhere between 1 in 100 and 1 in 1,000. So there will be one such hit for every 500 or so satellites. The probability of colliding with smaller pieces of debris is higher, simply because there are more of them. Between 100 and 1,000 pieces of debris less than one millimetre will hit the spacecraft over the same ten years.

NASA has developed a telescope installation specially for tracking objects in orbit. It is based in Cloudcroft, in southern New Mexico, at high altitude and far away from the glare of city lights. The telescope inside the main observatory dome is an example of a relatively new design: a liquid mirror telescope. Nearly all large telescopes, including the one at Cloudcroft, are reflecting telescopes, or reflectors. This type of telescope has a concave primary mirror – shaped like the dish of a satellite aerial. The primary mirror reflects light on to a smaller, secondary mirror, which in turn reflects it on to a light-sensitive electronic component called a CCD (charge coupled device). There is a CCD at the heart of every camcorder and every digital stills camera, which does the same job. When light forms an image on the light-sensitive elements of a CCD, it produces tiny electric charges, which can be amplified to produce a video signal. The signal can be displayed on a television screen or computer monitor, and can be recorded on videotape or as a digital 'movie' in a computer's memory.

Most reflectors have a primary mirror made of silvered polished glass. The primary mirror of the liquid mirror telescope is, as its name suggests, actually liquid: it is made of pure mercury. The mercury is poured into a 3-metre-diameter shallow concave dish, which rotates slowly and continuously – on a cushion of air to avoid shocks and vibrations. This causes the mercury to spread out into a thin, uniform coating on the dish, and it then makes a perfect primary mirror that is more reflective than a glass one. The telescope can point only straight upwards, otherwise the mercury would flow out of the dish. This restriction does not significantly inhibit the telescope's effectiveness, however, as most satellites and their associated debris travel in orbits that are inclined to the equator, and they all pass across the field of view once in a while. During each sighting, an object's speed, direction and distance are logged and analysed. The CCD produces an output that represents the field of view of the telescope.

The CCDs used in telescopes are extremely sensitive, and some can produce up to 200 images every second – eight times faster than in a camcorder. The sensitivity of the CCD in the liquid mirror telescope at Cloudcroft is one of the features that make the telescope useful for tracking tiny pieces of space junk. One of the astronomers who operates the liquid mirror, Mark Mulrooney, says, 'The mirror of the telescope is about 100,000 times larger than your eye, so it can collect about 100,000 times as much light as your eye; there's a CCD camera … about ten times as sensitive as your eye.' All this means that the liquid mirror telescope can detect objects that would have to shine about a million times brighter for an unaided eye to see them.

The display buzzes with activity, much of it 'noise' created by random signals produced in the CCD. However, the screen shows hundreds of stars, which appear as points of light drifting very slowly across the screen as a result of the Earth's rotation. Most of the stars on the display would be too dim to see with the naked eye, even on the clearest of nights. Every few seconds, a spot of light moves across the star field on the screen. Each of these is an object in orbit. From the speed of the spots of light, Mulrooney and his colleagues can estimate the object's altitude simply by looking at the screen. Looking at a slow-moving blob, Mulrooney comments: 'Ten thousand miles or so, about the size of a desk.' Many of the spots seem to flash on and off as they move across the screen. This indicates that the objects are tumbling. Satellites and pieces of space junk do not give out their own light: sunlight illuminates these objects, and the telescope picks up the reflected light. The flashing signal is due to the fact that the tumbling objects showing up on the display have shiny flat sides and flat solar panels. Mulrooney holds up a 2-centimetre-diameter ball bearing, and says that his telescope can detect such an object at an altitude of 1,600 kilometres.

The liquid mirror telescope was brought into service in October 1996. One of its first jobs was to track the 800 or so large fragments, and some of the countless small fragments, produced by the explosion of a commercial satellite in 1996. The satellite was launched in 1994, and its engine exploded in orbit at an altitude of about 600 kilometres. Information about the orbiting fragments, supplied by the liquid mirror telescope at Cloudcroft, was incorporated into NASA's flight plan for the Space Shuttle mission of February 1997. During that mission, the Shuttle had to orbit to about the same altitude as the cloud of debris from the destroyed satellite, to retrieve the Hubble Space Telescope for repairs. On inspection, the space telescope itself showed considerable damage from orbital debris. At one stage of the mission, the Shuttle orbiter, with the telescope in its bay, had to make a careful manoeuvre to a higher orbit, to avoid a book-sized fragment of the destroyed satellite that could have caused considerable damage.

The liquid mirror telescope is not the only tool that is used to chart and analyse space debris. Several powerful radar systems also monitor the orbiting debris. The radar stations are situated in defence establishments, and their primary function is to keep a watch on military satellites and the threat of incoming ballistic missiles. Increasingly, however, they are being called upon to look out for space debris. The information from radar installations is fed to the US Space Command, which co-habits, in the interior of a mountain in Colorado, with NORAD (North American Aerospace Defense Command). The door to this subterranean site weighs 25 tonnes, and would survive the blast of a hydrogen bomb. Inside the mountain, Space Command collates the information from the radar stations and supplies information to NASA and other space agencies. The radar can detect pieces of space junk even smaller than those observed by the liquid mirror telescope. Both types of installation are necessary, however: some objects reflect radio waves better than light, while others reflect light better than radio waves. Between them, radar and optical observatories can produce an accurate picture of the space debris environment for objects larger than a few centimetres.

In the UK there is another radar installation that tracks military satellites, and which is also increasingly used to track space junk. It is based at RAF Fylingdales on the Yorkshire Moors. The radar here is of a type known as solid-state phased array radar, or SSPAR. It does not have a spinning dish to transmit and receive radio waves, as most radar systems have. It is a huge eight-storey high pyramid with 2,560 small microwave transmitter-receivers arranged in a circle on each of its three faces. The total power output of the radar is equivalent to 25,000 100-watt light bulbs. Microwaves are very short-wavelength radio waves, including one with just the right frequency to agitate water molecules, increasing their temperature – this is why a microwave oven heats only foods that contain water. Any animal that came close to the powerful transmissions of the radar pyramid would be cooked within minutes.

The pyramid shape of the radar gives the team at RAF Fylingdales an all-round view of space, and they can detect orbiting objects that are just a few centimetres across and at an altitude of several thousand kilometres, anywhere above the horizon. Dave Tymon, an expert on orbital tracking who works at RAF Fylingdales, explains the importance of the sort of observations made there: 'We need to be able to protect certain satellites – especially those that are manned – to ensure that no piece of debris hits them; and to do that, we need to know accurately where each piece of debris is.' The ability of the radar at RAF Fylingdales to track and identify objects in orbit is quite amazing. When the French Cerise satellite went out of operation in 1996, its operators contacted RAF Fylingdales to see if the scientists there could work out what had happened. Dave Tymon recalls, 'We ran a simulation of Cerise against all the other pieces of debris and all other satellites over a five-day period.' Of the many thousands of objects in orbit, the team at Fylingdales managed to work out from their simulation that it was a piece of an Ariane rocket launched several years earlier that had hit Cerise.

Objects smaller than one centimetre inflict minor damage that causes degradation of orbiting spacecraft but does not generally threaten entire missions or astronauts' lives. Debris that is larger than a few centimetres can be tracked by the telescopes and radar observatories, and appropriate action can be taken in space. The real danger lies in the intermediate range – objects that are large enough to pose a real threat, but which cannot be detected from the ground. Richard Crowther of the Defence Evaluation Research Agency, in Farnborough, Hampshire, sums up the situation: 'With our radar and telescopes, we can see over 10,000 objects in space; of those, we've seen about twenty break up for unknown reasons, and we believe that they are due to collisions between those objects and objects we can't track.' As more collisions take place, more pieces of debris of intermediate size are created, and space becomes more and more hazardous.

What goes around comes around

You would think that the most obvious first step towards making space safer would be to send fewer objects into space. But there is no sign that the pace of spaceflight development is slowing: there are already more than a thousand satellite launches planned for the next ten years or so. And as hobbyists or teams registered for the X Prize (see 'Day Return to Space') thrust their rockets closer and closer to orbit, there may be even more.

Most of the planned launches are designed to place into orbit a new 'constellation' of satellites that will provide global telecommunications coverage. The first of these networks, Iridium, is already operational (although some of the more than seventy satellites already in orbit have failed and become nothing more than space junk). These satellites function in low Earth orbit, and make it possible for people anywhere on the planet to make telephone calls or connect to the Internet using a low-power transmitter. Until recently, this has been possible only by communicating via satellites in geostationary orbit, more than 35,900 kilometres away, which requires more powerful transmitters. Using satellites in low Earth orbit to communicate will also dramatically reduce the time delay experienced by current satellite communications. The next such network of low Earth orbit satellites is a system called Teledesic – a 300-satellite scheme backed by Microsoft's Bill Gates, which will eventually provide fast, cheap and global access to the Internet.

With more and more objects going into space, the amount of space junk in orbit will increase. However, as we have seen, space junk begets space junk: when one object collides with a spacecraft or discarded rocket fuel tank, it can cause an explosion or fragmentation that produces many more, smaller, pieces. Just as pebbles become sand as the sea crashes them together over thousands of years, some space scientists fear that the space junk currently in orbit will break down into smaller and smaller pieces, gradually producing a shroud around the Earth. This idea is called the Cascade theory, and was proposed by Don Kessler, who says that looking ahead, 'over a period of fifty years, you can have a significant increase in the fragment population'. He can foresee that population eventually forming a ring around the Earth, like those around Saturn. The objects that make up the ring would slowly coalesce, and may even form one huge artificial moon in low Earth orbit. This is the process by which, some astronomers believe, planetary satellites – including our own Moon – formed.

In 1990 space scientists Peter Eichler and Dietrich Rex at the Technical University of Braunschweig conducted a study of space junk. They used powerful computers to work out the likely scenario for the future of orbital debris. They showed that cascade really is possible, or even likely, and that 'critical mass' had already been reached. Since their research was conducted, hundreds of launches have taken place. Other scientists are more optimistic, claiming that the thousand or so planned satellite launches will turn out to be far fewer. They also point out that policies are now in place to reduce the debris-creating capabilities of new spacecraft. Several reports have been produced, with recommendations on how to reduce the impact of launch vehicles and defunct satellites on the space debris population. For example, parts of spacecraft that would normally be discarded – such as lens caps or straps that join the final stage of the rocket to its payload – will be attached to the spacecraft from now on, so that they cannot become space junk. And final-stage rockets themselves, most of which fall back to Earth after a few weeks or months, will be purposely propelled back, as soon as their payloads have been delivered into orbit. The incentive for commercial companies to spend money doing this, say the optimists, will be that in the long run they will save money, as fewer of their satellites will be put out of action by space junk.

But if the likes of Eichler and Rex are right, then the space community must take action to reduce the amount of orbital debris already in space, not simply send up less in the future. Some imaginative solutions have been proposed for doing this, notably by Kumar Ramohalli at the University of Arizona (whose work in robotics featured in 'What Shall we do with the Moon?'). As long ago as 1988, Ramohalli began work on an idea for an orbiting clean-up robot, called ASPOD (Autonomous Space Processor for Orbital Debris). ASPOD is designed to seek out satellites that have gone out of commission, and use a solar-powered laser to slice them into pieces. It would use robotic arms to recover reusable parts such as solar panels, and collect up any other pieces of space junk into a hopper. It would then re-enter the atmosphere, splash down into the sea and be recovered. Another plan that has been proposed to deal with existing space junk is Project Orion, suggested by Jonathan Campbell, a scientist at NASA's Marshall Space Flight Center. Campbell's idea is to use powerful ground-based lasers to knock pieces of debris out of orbit and into re-entry.

However, most space scientists – while they do not take lightly the dangers of space junk – are prepared simply to incorporate better shielding and more space-environment-friendly features into future space designs. The existing problem will go away, they say, as disused or fragmented spacecraft eventually fall back to Earth. The United States Space Surveillance Network has a catalogue of 23,000 space debris objects, of which about two-thirds have fallen back to Earth. According to Richard Crowther, 'Every day we observe at least one object coming back from space to Earth. These range in size from the order of a portable phone to large space stations.' Satellites at high altitudes – above a few thousand kilometres – can continue to orbit for hundreds, thousands or perhaps millions of years without any additional thrust. At these altitudes, there is virtually no atmosphere, so there is nothing to slow down the satellites' orbital speed. The atmosphere is very tenuous at about 200 kilometres, but it does exert a tiny amount of air resistance on objects in low Earth orbit. This slows them down, dragging them gradually lower, into more dense atmosphere, which slows them still further. Eventually, they enter the atmosphere proper and burn up. Space junk that burns up on re-entry appears as streaks of light, like meteor trails.

Meteors – or 'shooting stars' – are those fast-moving streaks of light produced when small pieces of rock from space enter the atmosphere. The pieces of rock (meteoroids) are mostly material left behind by the passage of comets around the Sun. Just as a high-speed collision of space junk with a spacecraft occurs where the orbits of the two objects are inclined with respect to each other, the orbits of comets around the Sun – and therefore of meteoroids – are steeply inclined to Earth's orbit around the Sun. This means that meteoroids approach Earth at great speed – sometimes as much as seventy kilometres per second. As it enters the atmosphere, a meteoroid collides with countless air molecules. This heats its surface – enough to vaporize it. If the meteoroid is small – less than about one millimetre in diameter – this process will continue until there is no more meteoroid left. The destruction of most meteoroids in this way takes only a second or two. The light that produces a meteor trail is actually created by charged particles in the vaporized region around the meteoroid, and not by the meteoroid simply glowing white hot. Larger meteoroids take longer to disintegrate, and leave longer, brighter trails. Meteoroids larger than about the size of a grape can actually make it to the ground without completely disintegrating. The centres of much larger ones can be found as chunks of rock – meteorites – and enormous ones can hit the ground with great force and cause damage. The main differences between space junk and meteoroids entering the atmosphere lie in their composition, shape and speed, but the underlying processes are the same.

Because space junk enters the atmosphere more slowly than meteoroids, its burn-up trails move more slowly, too. Meteors dart very quickly across the sky; a re-entry trail of a hunk of spacecraft looks like a slow-motion version. Size is important: as with meteoroids, small pieces of debris are more likely to disintegrate completely during re-entry than larger pieces. Large items of space junk, such as rocket fuel tanks, do not burn up completely. They can produce trails that last for several minutes. Most satellite bodies are made of aluminium, which is tough and machineable and has a low density, which means that spacecraft made of aluminium will be light for their size. Aluminium is worthy of being used to make space hardware for another reason: it has a relatively low melting and boiling point, so most aluminium objects that return from space burn up completely in the heat of re-entry.

Astronomer Paul Maley spends some of his spare time scanning the skies for space debris and, in particular, re-entry of space debris, using his binoculars. He is most proud of an event he managed to see and record on to videotape in 1984: 'Probably the most unusual and perhaps the most spectacular piece of debris was the re-entry of the Space Shuttle's external fuel tank that I observed from Hawaii … it began to break up into dozens of pieces that tumbled end-over-end and were flashing and flaring.' The re-entry of large objects like the Shuttle's external tanks creates an exciting fireworks display visible to the naked eye. Many UFO reports have been attributed to the re-entry of space junk into the atmosphere. Some spacecraft are equipped with 'debooster' rockets that can slow them down at the right moment to descend into the ocean, in a controlled re-entry. This is likely to become more commonplace. NASA's 1997 Policy for Reducing Orbital Debris Generation stresses the importance of 'post-mission disposal', for example.

Space agencies can normally predict with some accuracy where objects will land, although this task is more difficult for objects with irregular shapes. The larger or more durable objects that re-enter and make it down to ground level nearly all end up in the sea, since about two-thirds of the Earth's surface is ocean. Occasionally, large pieces of space junk that survive re-entry do hit the ground. One such object was NASA's first space station, Skylab. Launched into orbit in May 1973, Skylab was in use until February 1974 as an orbiting science laboratory, and then was moved into an orbit that would keep it in space until 1983, when the Space Shuttle programme was to begin. However, increased solar activity upset NASA spaceflight engineers' calculations of the air resistance of the upper atmosphere, which began to drag Skylab down noticeably in 1978. More than a year later, in July 1979, the craft was hurtling down to the ground, heading for Australia. The people of Perth, in Western Australia, seemed to be the most at risk, but in the end the surviving fragments of the former space station were strewn across a large uninhabited area.

Another large piece of space debris to make landfall was a fuel tank, with a mass of 260 kilograms, in January 1997. The fuel tank hit Georgetown, Texas, and landed just 40 metres from farmer Steve Gutowski's house. Gutowski recounts the events of that unusual day: 'My wife went out to get the newspaper, she says, "You'd better look out there, boy, there's a dead rhinoceros out there" ... then I looked up to the sky and I said, "There's only one place it could have come down from: heaven."' The fuel tank – from a Delta II rocket – was cylindrical, about three metres long and 1.5 metres high, and had been crushed a little and broken open due to the re-entry and impact. The tank was taken to the Center for Orbital and Re-entry Debris Studies, part of the Aerospace Corporation, in Los Angeles. There it joined a collection of other ex-orbiting objects.

Director of the Aerospace Corporation, Bill Ailor, described the tank: 'You notice this quite prominent hole – we don't know exactly what caused that hole, but it looked like – based on the shape of the metal, the fact that we've seen some high-temperature process going on here – this probably occurred relatively early in the re-entry.' Another of the objects in Bill Ailor's 'space junk museum' was found about thirty kilometres away from Steve Gutowski's farm, by an Airedale terrier called Oliver. What Oliver came across was a beach-ball-sized sphere of titanium. This, too, was part of the Delta II rocket. It once contained pressurized gas that was used to force fuel into the rocket engines. There were three more identical spheres on the spacecraft, and they are probably lying undiscovered in the Texas countryside.

In neighbouring Oklahoma on the same day, Lottie Williams was hit by a small piece of debris, from the same spacecraft, while walking in the park: 'My first instinct was that it was something from a shooting star.' Ms Williams is the only person to date to have been hit by re-entered space junk.

At the Center for Orbital and Re-entry Debris Studies, Ailor and his colleagues analyse samples in a mass spectrometer, which enables them to determine the mixture of chemical elements present. Ailor claims that he can often tell where a satellite was made from the fragments they have recovered, since space agencies in different countries tend to use different alloys. Analyses of re-entry fragments are useful to aerospace engineers, who can check their understanding of the materials they use in the construction of satellites and launch vehicles, and ensure that they do burn up in the atmosphere, or land in a predictable location so that accidents can be avoided.

Break-up of the nuclear family

Re-entry may be useful in reducing the amount of junk in orbit, but it can sometimes create a different problem. More than seventy of the satellites launched into space have carried radioactive material, which can be spread over a large area if the satellites re-enter. In 1970 the first of a series of Russian navigation satellites was launched. These were RORSATs (Radar Oceanic Reconnaissance Satellites), which used radar to track ships at sea. Solar cells would not produce energy fast enough to provide power for the radar, so the satellites carried nuclear generators, each holding about thirty kilograms of uranium-235 fuel. Of the planned thirty-three nuclear-powered RORSATs, thirty-one made it into orbit. All but two of these were moved into 'storage orbits', at around 1,000 kilometres, where they will remain for at least 200 years. Spent nuclear fuel on Earth is normally encased in thick concrete and buried deep underground. Either way, the fuel is highly radioactive, and will not be reasonably safe for thousands of years.

One of the RORSATs that did not make it into the safe storage orbit was Kosmos 954. Launched in September 1977, Kosmos 954 re-entered the Earth's atmosphere – with its nuclear reactor – in January 1978. It broke up and, despite features designed to keep each reactor element in one piece, scattered millions of tiny radioactive pieces over Canada's Northwestern Territory. The scattered material amounted to about one half of the reactor core – the remainder of the core remained intact. Most of the uranium was recovered by a team from the CIA (the US Central Intelligence Agency); some of the pieces recovered from the ground were as small as sugar granules, and the clear-up operation cost $13 million. Four years later, a similar incident with another nuclear-powered Kosmos satellite sent the 30 kilograms of uranium into the Atlantic Ocean. It is the RORSATs still in orbit that are leaking sodium and potassium coolant into space, too.

Another nuclear-powered Russian spacecraft, Mars-96, was at the centre of controversy in November 1996. The craft was destined for Mars but, just one day after launch the final stage failed and it re-entered the atmosphere. According to both the Russian space agency (RKA), and the US Space Command that was tracking the craft, Mars-96 plummeted, its load of plutonium intact, into the sea just off Chile. However, eyewitnesses report seeing a fireball break up into pieces over the Atacama Desert. Patricio Aravena, a Chilean policeman, was one of those who saw the break-up: 'It was ten o'clock at night; I was standing here watering the yard when a bright light appeared in the west. It was very strong … as it came towards us, it broke into hundreds of bright lights that were still glowing as they vanished over the horizon.' Space Command were not the only ones who were tracking the trajectory of Mars-96. Astronomer Luis Barrera says that he was 'able to plot the trajectory of the rocket carrying the probe; the rocket did explode near to the coast, where some of it landed. But the probe itself carried on to the mountains.' No pieces of debris were found, but the Atacama is a vast, sparsely populated area.

It is not only the Russians who have sent up radioactive materials into space. NASA has used plutonium-238 in electrical generators in about thirty missions. Most of these missions have involved space probes that were to travel to planets farther away from the Sun, where the intensity of sunlight is far less than in the vicinity of Earth. The plutonium is used as the source of power in a 'radioisotope thermoelectric generator', or RTG, which uses a junction of two metals to generate electricity from heat. A small RTG is used in some heart pacemakers, instead of batteries that would have to be replaced. The plutonium in an RTG is in the form of solid pellets of plutonium dioxide. Unlike the Russian nuclear power generator on the RORSAT satellites, an RTG does not use a fission chain reaction to generate heat. Instead, it produces heat by the natural decay of radioactive plutonium. During decay, the nuclei of atoms of plutonium-238 release high-speed electrons – a process called beta radiation. The electrons bump into neighbouring atoms, and this is what heats up the plutonium dioxide. The rate at which the heat is released is just right to power the spacecraft.

There is great concern among many people about the use of plutonium in RTGs. Plutonium-238 does not occur naturally on Earth: it is produced in nuclear reactors. It is poisonous and remains radioactive for many years. In tests, tiny amounts of plutonium inhaled by animals led to cancer. Just a single decaying atom of plutonium inside a living cell can disrupt the DNA inside, and lead to abnormalities in a newborn child. Having said all that, the conker-sized plutonium dioxide pellets in NASA's RTGs are encased in extremely durable iridium and graphite; if they do break, they break into large pieces, not into pieces small enough to be inhaled; and they are extremely heat-resistant and will not vaporize during re-entry or a fire. All these features are designed to reduce the chance of a release of plutonium dioxide. The safety of RTGs was brought up to this standard after an accident during the launch of an American navigation satellite, SNAP-9A, in 1964. The satellite employed a type of RTG called a SNAP (system for nuclear auxiliary power). The satellite blew up over the Indian Ocean and released about one kilogram of plutonium-238 into the atmosphere. The radiation 'footprint' of this accident can still be measured – once the plutonium-238 has released beta radiation, the product of the decay goes on to produce alpha radiation, which is more dangerous.

The largest amount of plutonium ever sent into space is being carried by the Cassini space probe, launched on 15 October 1997 and due to arrive at Saturn in June 2004. It will go into orbit around the planet and, in November 2004, it will drop a small probe on to the surface of Titan, Saturn's largest moon. Saturn orbits the Sun at an average distance of more than 1,420 million kilometres, where the intensity of sunlight is about one-hundredth of that near Earth. So aboard the probe are three RTGs containing a total of 32.4 kilograms of plutonium-238. This mission has been accompanied from the outset by tremendous controversy and has provoked widespread protests. Nonetheless, NASA is confident that its RTGs would survive re-entry and would not scatter plutonium in the atmosphere, and they have another eight missions planned that may use RTGs. Some space probes use radioactive plutonium to keep their instruments warm, rather than to produce electricity. The RHUs (radioisotope heater units) have been used on several previous missions, and NASA plans to use them in space probes that they are considering for missions to Mars and Jupiter and a rendezvous with a comet.

With so much other space junk in orbit, which may cascade into smaller and more numerous fragments, many people think that sending radioactive substances into space is foolhardy. Michio Kaku, a nuclear physicist at the City University of New York, is a prominent voice in the movement against the use of nuclear power in space: 'It's only a matter of time before one of our space probes encounters some kind of space debris in space. I say it's like Russian roulette: sooner or later, you press the trigger often enough, one of our space probes will in fact encounter space debris and spew plutonium on planet Earth.' Whatever the dangers associated with large objects either in orbit or falling back to Earth, space engineers are now engaged in building the largest manufactured object ever to orbit the Earth: the International Space Station.

Action station

By 2004 the ISS (International Space Station) will be up and running. It is one of the most ambitious scientific projects of all time, involving an unprecedented level of international collaboration. The project is being co-ordinated by NASA in the USA, but Russia, Brazil, Canada, Japan and eleven member countries of the European Space Agency are also involved. When fully assembled, the ISS will be 109 metres long and 88 metres wide, and will have a mass of 500 tonnes. It will be four times the size of the current largest orbiting object, the Mir space station. The ISS will consist of modules, connected together in space, and powered by about 4,000 square metres of solar panels that can be turned to face the Sun. It will orbit at an average altitude of 400 kilometres, about the same as the altitude of Mir. Scientists will spend between three and six months there, carrying out a range of experiments in near-zero gravity ('microgravity'). When complete, the ISS will be visible to the naked eye, by virtue of light reflected off its surfaces. In fact, it will be one of the brightest objects in the night sky – brighter than any star and all the planets except Venus.

For such a large object, in low Earth orbit, the chances of being hit by large pieces of orbital debris are significant. In June 1999 a spent Russian rocket came within one kilometre of the two modules that were in place at that time. space-walks around the exterior of the ISS will be routine – more than 160 of them will be conducted during the construction programme alone – and this will increase the risk to the astronauts. In order to be able to calculate the chances of the ISS being hit by debris, NASA put together a test package that members of a Space Shuttle crew attached to Mir in March 1996. The package – called the Mir Environmental Effects Payload – consisted of four briefcase-sized units. One of the units contained an alloy plate on to which small meteoroids and small pieces of space debris impacted. The package was exposed to the space environment until September 1997, when it was retrieved by another Space Shuttle mission. On inspection under a microscope, thirty-eight tiny craters were found, which had been made by hyper-velocity impacts of very small particles. The particles vaporized on impact, and so it is difficult to tell much more than their mass and speed. One of the other units used in the experiment was cushioned, so that it could absorb the impact, and collect the kind of tiny pieces that produced the craters in the first unit. Once this unit was returned to Earth, analysis showed that there was a mixture of micrometeoroids and space junk. The space junk included tiny paint flecks, powdered solid rocket fuel and tiny pieces of aluminium.

The Mir Environmental Effects Packages allowed researchers to work out what kind of dangers the ISS may face, and what kind of shielding it may need. At the Hypervelocity Laboratory in the White Sands Testing Facility in New Mexico, they determine not only what type of shielding to use, but also where it is most needed. Using knowledge of the most likely direction of approach of orbital debris, together with the direction in which the ISS will be facing in orbit, researchers at the laboratory are able to construct computer models of the space station showing the risk of impact. The high-risk, vulnerable surfaces are shown in red, the low-risk, least vulnerable in blue. Justin Kerr, one of the shielding designers for the ISS, says 'If you're a shielding designer for the International Space Station, you want to put your most robust shielding in the front of the spacecraft, in the red areas.'

If the ISS is hit by a large object, it could be a catastrophic collision. The knowledge of the distribution of space junk in orbit will help to give adequate warning, but the ISS is less manoeuvrable than, say, the Space Shuttle. In case of an emergency, there is an escape vehicle for the astronauts – the equivalent of a lifeboat. The experimental version of this vehicle is called the X-38, and has already undergone extensive tests. It is a wingless, lifting body, similar to the X-33 and VentureStar spacecraft described in

'Day Return to Space', but much smaller. It would be deployed quickly and efficiently if an emergency occurs. This vision of escape pods leaving space stations is like something from a science-fiction film. But as long as space remains full of junk, escape may be necessary. If the ISS is hit by a large object, a lot more debris will be created. In the worst case scenario, that might start off the cascade that will fill our skies with tiny pieces of debris. Alternatively, the space station may fall back to Earth. But if it does, will we be able to predict where it will fall?

SUN STORM
... searching for a space
weather forecaster...

On a clear night, with the unaided eye, you can see about 6,000 stars. On a clear day, you can see just one – the Sun. Like any of the stars visible at night, the Sun is a huge nuclear furnace radiating enormous amounts of matter and energy into space. The Sun is extremely large – 109 times the diameter of the Earth – and it contains more than 99 per cent of the total mass of the Solar System. And it produces vast amounts of energy, which radiates out in all directions. The Sun is 150 million kilometres away from us – to give you an idea of just how far away that is, consider how long it takes for sunlight to reach us. Light travels 1,000 kilometres through space in just one three-hundredth of a second, and yet sunlight now arriving at the Earth left the Sun a full eight minutes ago.

Despite the immense distance between the Sun and the Earth, life on Earth depends upon the Sun absolutely. Even a small change in the Sun's output of light has dramatic effects on the Earth. But we do not receive only light from the Sun. The 'surface' (outer area of the ball of gases) of the Sun is a seething, turbulent place, and it emits a stream of energetic particles across vast distances. Sometimes, immense blobs of searingly hot gas are ejected, too, during periods of disturbance in the Sun's magnetic field. This is what some people refer to as a sun storm, and the effects of this 'space weather' phenomenon may be felt on Earth. Some of the effects of a sun storm are beautiful: the auroras, such as the northern lights, are much more intense when a sun storm hits. However, there are plenty of undesirable effects too, such as widespread power cuts, satellite malfunctions and hazards to the health of passengers in aeroplanes at high altitude. And the more we have come to rely on technology in our lives, the more of a threat sun storms have become.

Great ball of fire

One of the popular images of the Sun is that it is a huge ball of fire. In one sense, it is. Both fire and the Sun are made of hot gases, which we can see because their high temperature makes them glow. However, that is where the analogy ends. In a fire, the hot gases are released as a result of a chemical reaction. When a candle burns, for example, carbon and hydrogen in the candle wax combine with oxygen from the air to make carbon dioxide and water, and release energy as heat. The products of this reaction – the carbon dioxide and the water – are gases, and are produced at the wick. It is because these gases are hot that they produce light – a process called incandescence. The surface of the Sun is also incandescent, but, in this case, the heat is generated by nuclear reactions, not chemical reactions.

The Sun consists mainly of two gases, hydrogen and helium. About 75 per cent of the Sun, by mass, is hydrogen, and it is hydrogen that takes part in the fusion reactions that power the Sun. When the Sun was born, about 5,000 million years ago, it was a cloud almost exclusively made of hydrogen. Gravity caused this cloud of hydrogen to pull in on itself until the centre of the Sun became dense and cramped, heating the hydrogen gas there. At the high temperatures in the young Sun, the electrons were stripped from their atoms, leaving bare hydrogen nuclei. A gas in which electrons have been separated from their atoms in this way is called a plasma. The contraction of the plasma cloud of hydrogen would have continued had it not been for what happened next.

When the temperature at the centre of the young Sun became high enough, the hydrogen nuclei slammed against each other hard enough to join, or fuse, producing nuclei of helium. This nuclear fusion released huge amounts of heat, and continues to do so. Every second, about 700 million tonnes of hydrogen are converted into about 695 million tonnes of helium. The missing mass is carried away as energy – it was Albert Einstein who first realized the bizarre fact that mass and energy are inter-changeable, as summed up by his famous equation $E = mc^2$. The 'E' represents energy, the 'm' represents mass and the 'c' is the speed of light (through empty space). It works out that the loss of 5 million tonnes every second at the Sun's core liberates as much energy as it would take to light 4,000 million million million 100-watt light bulbs. The Sun's energy output at the surface – mainly as visible light and infrared – is equivalent to having 630,000 100-watt light bulbs in each square metre of its surface: no wonder the Sun looks bright. The nuclear fusion releases its energy as gamma rays that heat the Sun's core. The temperature causes the core to expand, fighting against the gravita-tional collapse. The expansion and contraction are in balance, and the Sun produces heat at a fairly stable rate as it continues to fuse hydrogen. Eventually, the balance will be destroyed – when the Sun begins to run out of its hydro-gen fuel. This will be a very long way into the future, and will result in dramatic changes to the Sun, which are discussed more fully in Chapter Five.

Deep in the Sun's core, the temperature is thought to be in the order of 15 million degrees C. At this temperature, the gases in the core radiate huge amounts of energy, largely as gamma rays and X-rays. The intense radiation heats the surrounding layers, which re-radiate the energy to the layers above them. At each successive layer, the temperature is less than the one beneath it – this is because the heat is being transferred outwards in all directions, becoming more spread out in space. The heat transfer by radiation continues outwards. Towards the surface of the Sun, however, the main method of heat transfer is convection. It is convection that is behind the distribution of heat in a pan of water on a stove. The pan is heated at the bottom, and the temperature of the water there increases. This water becomes less dense as it heats up, and it floats to the surface – pushed upwards by colder, more dense water intruding from above. When the water reaches the surface, it cools, and is pushed aside by the water that had replaced it at the bottom of the pan. As it cools still further, it sinks to the bottom again. This churning is called a convection cycle, and it helps to spread heat throughout the water. Due to convection, a pan of water on a stove can be a turbulent place. The same is true of the gas in the outer layers of the Sun. In all, energy generated in the core takes 50 million years to reach the surface by radiation and convection. Even if the nuclear reaction at the core stopped, the Sun would continue shining for many millions of years.

The surface of the Sun is of course vastly more turbulent than a pan of boiling water, as can be seen by the beautiful images produced using large telescopes. These images show how the convection of gases at the surface forms 'cells', each with hot material rising to the surface in the centre and cooler material sinking around it. Together, the cells give the Sun a granulated appearance in these images. The turbulent, granulated appearance of the Sun is due to the twists and turns of its intense magnetic field, which is very different from that of the Earth. Most people are familiar with the shape of the Earth's magnetic field, which is similar to the field of an ordinary magnet. Iron filings sprinkled around a magnet on a tabletop map out the magnet's field. They form lines that loop from one end, or pole, of the magnet to the other. These lines represent the direction of magnetic forces, which align the iron filings between the magnet's north and south poles. The magnet's field is a representation of the direction and strength of these forces. The Earth's magnetic field is very similar to the field around a bar magnet: it, too, loops between the north and south poles. The Sun's magnetic field, on the other hand, is far more complex than the Earth's and is constantly shifting. Sun storms are related to the complex shifting loops of magnetic field lines at and above the Sun's surface.

During a sun storm a huge mass of gas – up to a thousand million tonnes – is ejected from the outermost region of the Sun, the corona. The gases that make up the corona are essentially the same as those making up the rest of the Sun, but for reasons that are not fully understood they are much less dense and far hotter than those at the surface. The temperature at the surface is about 5,500 degrees C, while the temperature in the corona is more than two million degrees C. At this temperature, electrons in all the atoms are stripped from their nuclei – the corona is an extremely hot plasma. Energetic particles from this plasma – mainly high-speed electrons and hydrogen nuclei – are ejected at high enough speeds to escape the Sun's gravity, forming a constant stream from the corona called the solar wind. It has an average speed of about 400 kilometres per second. During a sun storm, unlike the constant stream of the solar wind, plasma is thrown off the corona in individual huge blobs called coronal mass ejections. These blobs are as much as forty times as dense as the solar wind. The frequency of sun storms varies with the activity of the Sun as a whole. During times when the Sun is active, there may be a few ejections every day, while at other times there may be only one every week or so. As they leave the Sun's corona, these ejections expand slightly, and travel out into space.

The coronal mass ejections produced by sun storms travel away from the Sun at extremely high speed. Sometimes they are directed towards the Earth, and take between three and five days to reach us. When a coronal mass ejection meets the Earth, it can produce electrical and magnetic disturbances. A heavy sun storm affects magnetic compasses all over the world, for example. When solar plasma reaches the Earth, it follows the lines of the Earth's magnetic field, because it consists of electrically charged particles. The field lines converge at the Earth's magnetic poles, which lie near to the geographical north and south poles. The electrons and hydrogen nuclei of the plasma travel at incredibly high speeds and, when they enter the atmosphere around the poles, they slam into air molecules, producing an eerie but beautiful glow. This eerie light often takes the form of a waving green-and-red curtain – an aurora. Around the northern hemisphere, the aurora is called the northern lights, or aurora borealis. Around the south pole, it is the southern lights, or the aurora australis. The air molecules become 'excited' – they have more energy – and when they release their excess energy, it is in the form of light.

Dr Nicola Fox, who works at NASA's Goddard Space Flight Center in Maryland, has carried out extensive studies of auroras. She says that the phenomenon is 'completely spellbinding'. There is a faint aurora all year round in the polar regions, because there are always solar particles arriving at the Earth courtesy of the solar wind. But when a sun storm hits, the intensity – and the beauty – increases. Fox explains the different colours present in auroras: 'If you see a lot of green you're seeing a particular excited state of oxygen; if you see a lot of red, you're seeing a different excited state of oxygen.' The green light is more predominant when the particles entering the atmosphere have higher energies than normal – when they have been produced by a sun storm. However, sun storms do more than create beautiful light shows. According to Fox, violent solar events 'can send bubbles of hot plasma streaming from the Sun to the Earth, and here it can cause chaos'.

Solar power cut

At 3.45 am on 13 March 1989, Marie-Claude Bertrand was at home in Quebec: 'All of a sudden, everything just blacked out,' she recalls. She was not the only one whose lights failed that morning, however. Power was lost to about eight million homes, and all Quebec's businesses. Another resident of Quebec, Michael Bailey, looked outside his window when the power cut hit: 'We looked at the city ... we could see different sectors just switching off.'

The power cut was the result of a severe overload on the city's entire electrical supply system. Night workers at the Quebec National Grid Control Center had noticed the surge, and had done their best to cope with it. But the electric current was too powerful, and it knocked out power stations and substations across the grid. It took eight days to get the system back to normal. The surge of current that debilitated Quebec's electricity supply grid was caused by the arrival on Earth of a coronal mass ejection, which had been thrown off the Sun on 6 March. The fast-moving electrically charged particles arriving on Earth induced powerful electric currents in the grid – the energy of the stream of plasma would have been enough to vaporize the Mediterranean Sea. Unfortunately for the residents of Quebec, their city lies on a bed of granite, which is not a good conductor of electricity. If Quebec had been in a different geological location, the ground might have taken the brunt of the electric shock.

During this great geomagnetic storm, other electrical distribution systems were affected across Canada and the USA. A huge transformer was damaged beyond repair in a nuclear power plant in New Jersey, for example. Power cuts were experienced in Sweden, too, as the result of this sun storm, and auroras were observed in regions much farther from the poles than normal – the northern lights were observed as low as New York and London. In assessing the storm, the Oak Ridge National Laboratory estimated that a similar storm in the north-east of the USA could cause damage worth as much as $6,000 million.

The people of Canada were affected by another sun storm five years later, in 1994. This time, two key communications satellites were put out of action. The first satellite went out in the late afternoon. Bruce Burlton is an engineer at Telesat, the company that was operating the satellites. He explains that the satellite 'had spun out of control – it wasn't looking down at the Earth so it couldn't carry communications channels'. Television pictures were lost across much of America and Canada, telephone conversations were cut off, and aeroplanes at several airports were grounded as a result of the loss of the satellite. After five hours of painstaking work, the engineers managed to recover the satellite, stabilize it and return the communications to normal.

However, the disruption was not over: another of Telesat's satellites was put out of action a few hours later. Len Lawson, Telesat's marketing director, had just sat down to relax and watch television when he noticed that the set was not receiving any signals: 'The sports network had gone to noise ... the music station was gone to noise ... the local TV station was gone to noise.' Like the first satellite, this one had also spun out of control – but this time it took five months to recover it. The loss of either satellite would have cost $100 million, which would have meant bankruptcy for the company.

It was a coronal mass ejection from the Sun's corona that had caused the loss of both satellites. The electrons contained within the plasma were absorbed by insulating materials on the satellites, such as thermal blankets that keep the spacecraft warm. When electrons build up and cannot leak away, an electric charge accumulates. This can give rise to an electrical discharge, which can damage a satellite's electronic circuits. Several other satellites have been put out of action in this way. In 1997 AT&T's Telstar 401 satellite was lost during heightened solar activity, cutting transmission on several television channels. And in 1998 more than forty-five million pagers in the USA became inaccessible when the satellite upon which they depended was lost due to a sun storm. All these satellites are in geostationary orbits, at an altitude of about 36,900 kilometres, and are more vulnerable to the effects of sun storms than satellites in low Earth orbit. Those in lower orbits are protected by the Earth's magnetic field, which diverts the plasma away rather like the bow of a boat pushes water aside. As it does so, it produces a teardrop-shaped region of space called the magnetosphere, cushioning us from the blowing solar wind. The magnetosphere is our very own solar windbreak, and it changes shape with the strength of the solar wind.

While sun storms are most likely to knock out the electronic circuits of satellites in geostationary orbits, they can affect satellites in low Earth orbit in a different way. The energy of a cloud of solar plasma released by a sun storm actually causes the Earth's atmosphere to expand slightly when it hits. This increases the drag on satellites in low Earth orbit, which slows them down. When this happens, people on the ground whose responsibility it is to track satellites lose them temporarily. The main organization that tracks satellites in the USA is Space Command at NORAD (North American Aerospace Defense Command) in Colorado. 'When we have a really bad sun storm, we start losing track of the satellites because they're not where they should be,' says Major Gregory Boyette of NORAD. 'During the last serious one I can recall, it took approximately ninety-six hours for us to reacquire our satellites, and figure out exactly where they were, so that we could continue tracking them.' Dr Daniel Baker of the University of Colorado thinks that this could actually be a threat to national security: 'If a country is intent on military mischief and they are sophisticated enough to understand that solar disturbances can have a detrimental effect on our knowledge of what's going on in space, then the worst-case scenario is that they would choose that time to launch some kind of a strike or some kind of military action.'

The expansion of the Earth's atmosphere caused by the arrival of a huge blob of solar plasma can also be helpful. It can cause orbital debris – space junk – to fall to Earth, burning up as it re-enters the atmosphere. During a sun-stormy year in 1991, an estimated 2,000 sizeable items of space junk fell to Earth.

Violent solar activity has also been shown to affect long metal pipelines. Powerful electric currents are induced in these pipelines, which accelerate corrosion. Sun storms can affect the health not only of satellites and pipelines, but of people, too. Astronauts working in space are exposed to the plasma from coronal mass ejections – particularly if they orbit over the poles, where the solar plasma does move close to the Earth because the magnetic field lines converge there. The effects of exposure to high-energy plasma from the Sun are the same as a high dose of radiation: the electrons and hydrogen nuclei can pass into living cells and disrupt the DNA inside. This can lead to cancer. When astronauts are inside a space capsule, they are protected to a certain degree from the flow of plasma. During a spacewalk, however, a spacesuit alone is little protection against extremely fast-moving charged particles. And several Russian astronauts who have conducted spacewalks have since developed cancer and radiation sickness, although it is not certain that the solar wind is to blame. Still, Professor Mike Lockwood at the Rutherford Appleton Laboratory in Didcot, Oxfordshire, says that he 'personally wouldn't be terribly happy about being inside a spacecraft that flew over the auroral regions on a regular basis'. Perhaps we should all think twice about day trips into space and hotels on the Moon after all.

But even on Earth we may not be safe from the Sun's energetic output. Charged particles from the Sun can affect people in high-altitude aircraft, such as Concorde, which cruises at an altitude of 18.5 kilometres. Since it began operating in 1976, Concorde has carried on board a device that detects solar plasma particles. If there is an unusually high flow of the particles, then the pilot drops the aeroplane to a safer altitude. According to NASA's Nicola Fox, 'You usually find the airline companies will switch around the crews on a much more regular basis when we have a violent [solar] event.' The effects of sun storms may even cause health problems at ground level: a doctor in Israel has conducted a study into a possible link between solar activity and heart attacks, the main consequence of which is myocardial infarction. Professor Eliyahu Stroupel, of the Bellinson Medical Centre in Tel Aviv, says that his data showed that 'in years of high solar activity, we have more deaths from myocardial infarction, especially in the older population'. A study in Hungary has even suggested that there may be more car accidents when the Sun is more active.

As increased activity on the Sun can have so many effects here on Earth, it would be advantageous if space scientists could predict when it was about to occur. Professor Mike Lockwood says of the space science community: 'We're very interested in what is going on inside and on the surface of the Sun, because that tells us about the likelihood of events that can wipe out our modern high-tech systems, on which we have become very dependent.' A reliable forecast of the 'space weather' would be very useful to many people beyond the space community, too. The effects of solar activity on communications and electrical power systems on Earth have been known for decades. Since the early 1980s an organization called the Space Environment Center, in Boulder, Colorado, has been issuing warnings to airlines, satellite operators and power companies during times of extreme solar activity. But its predictions of the effects on Earth were often wrong. In fact it was wrong two-thirds of the time (reversing the forecast each time would have increased its accuracy by 33 per cent). During the 1990s there was a great deal of activity aimed at improving the predictions. In 1994, for example, several agencies in the USA got together to consider setting up a co-ordinated effort to improve the forecasting of space weather: the National Space Weather Service. At the same time, there was a growing interest in the Sun in the scientific community. An intercontinental initiative called the International Solar-Terrestrial Program involved proposals for a number of space probes that would increase our understanding of the Sun. This is the latest, very high-tech development in solar investigation – which itself has a long history.

Seeing the light

Today, astronomers have a detailed knowledge of the Sun, although there are still many mysteries still to be solved. Before the birth of modern physics, people had no hope of actually understanding how the Sun provides the heat and light upon which we depend. They could really only notice how it moves across the sky, and wonder what this hot, bright ball might be that rose in the east and set in the west every day without fail. Many ancient cultures held the Sun as the focus of their spiritual celebrations, they built temples or monuments to help them in their sun-worship. Stonehenge seems to be a 5,000-year-old solar observatory, and there are many similar structures throughout the world. The position of the Sun against the stars at the time of a person's birth was thought to be a major influence in that person's life, and many people still believe this to be true. Despite the importance people placed on the Sun, it took a great deal to convince some people that the Earth moves around it, and not vice versa. For example, Galileo was imprisoned by the Catholic church and forced to deny his claim that the Earth moves around the Sun.

The Sun was important to the ancients in many practical ways, as well as spiritual or ideological ones. Its movements across the sky over the period of a year were the basis of the calendar, and it was used to measure the time of day, by observing the shadows it cast on sundials. The Sun was used in navigation, too: by simply measuring the position of the Sun in the sky from a particular location, it is possible to tell that location's latitude. In the sixteenth century, sailors used a cross-staff to make more accurate measurements of the Sun's position. The cross-staff was a stick about a metre long, with a movable cross-piece. A sailor would look along the length of the stick, and point it at the horizon. Then he would move the cross-piece until the top of it was aligned with the Sun (or, at night, the pole star). Many sailors were blinded as a result of their repeated use of the cross-staff.

The first truly scientific investigations into what the Sun might be had to wait until the invention of the telescope. The first astronomical observations using telescopes were made at the beginning of the seventeenth century. It was learned early on that it was not a good idea to look directly at the Sun through a telescope, and soon a safer way to view it was discovered using the magnifying power of a telescope. This involves projecting the visible face (the disc) of the Sun on to a screen behind the telescope's eyepiece. Galileo was among the first to do this. The Sun's disc is still projected in this way today, in large solar observatories as well as by amateurs using small telescopes or even binoculars.

One of the features that is immediately noticeable in a projected image of the Sun is the presence of groups of small dark patches, called sunspots. The early astronomers argued about whether these sunspots are actually on the

Sun's surface – some believed that the Sun, being a celestial body, could not be 'blemished' in any way, and so the dark patches they could see were somehow shadows of something above the Sun's surface. It is now known that sunspots are, in fact, on the Sun's surface. Sunspots were actually noted long before the invention of the telescope, by ancient Greek astronomers – they could see them when they gazed at the Sun through thick clouds or fog. In the first century BC, astronomers in China used the same method to chart the way these dark patches shifted and changed. These observations are far simpler and more consistent when noted through a telescope, and detailed drawings of sunspots were produced by many people from the seventeenth century on. These drawings highlighted how the sunspots moved from day to day, across the face of the Sun. Because sunspots always form in groups, you can follow a particular group as it moves across the Sun's disc. In fact, if groups of sunspots persist long enough, they can be seen disappearing from one side of the disc and reappearing at the other side two weeks or so later. This is because the Sun is rotating, just as the planets are.

Sunspots normally have a lifetime of several days or weeks, and the population of sunspots present varies over time. In fact, the number of sunspots at any time is an important indicator of solar activity. When there are more sunspots, there tend to be more auroras, for example. The number of sunspots, and therefore the Sun's activity, varies over a repeating period of about eleven years. This 'solar cycle' was first discovered in 1843, by the German astronomer Heinrich Schwabe. The solar cycles have been numbered, starting in the 1750s when reliable records of sunspot numbers were first made. The current solar cycle is number 23, and it runs from 1995 to 2006, with a peak in

2000. The last solar cycle was at its maximum at the end of the 1980s, when the violent sun storm hit Quebec's power system. At that time, there was a particularly large group of sunspots – twenty-seven times the average size. There was also an increase in solar flares – another feature of the Sun's turbulent surface, also related to the complex twists and turns of the solar magnetic field.

A solar flare is an eruption of material from the surface of the Sun, which normally takes place along the dividing line between two sunspots. Coronal mass ejections – the really big solar events – are often related to both sunspots and solar flares, but they can happen independently of them, too. The material ejected during one of these violent events comes from the corona, not the surface of the Sun as with solar flares. Sun storms happen all the time – just more frequently and more energetically around the maximum of solar activity. Many people predicted that all the communications satellites in orbit would be knocked out by cataclysmic events at the solar maximum at the beginning of this new millennium – but to date there have been no such disasters.

Not only does the sunspot cycle generally correspond to a variation in the frequency of auroras; it seems to be intimately linked with our weather and climate here on Earth. In the 1890s British astronomer Walter Maunder noticed from historical records of sunspot observations that there were hardly any reported from about 1645 to 1715. This seventy-year period corresponded to a long cold spell, sometimes referred to as the Little Ice Age. During this time, the weather was so cold that rivers froze solid, and in Europe many cities held 'frost fairs' on them. There was another particularly cold period during the first two decades of the nineteenth century; again, this corresponded to a time when solar activity was low. Of course, the climate depends upon so many factors that it is difficult to say for certain that a reduction in the solar activity – indicated by the reduction in the number of sunspots – is necessarily the cause of mini-ice ages; it could be mere coincidence. However, more recently, similar evidence has come to light.

When John Butler joined the Armagh Observatory in Northern Ireland, he was interested in the history of observations made there. The observatory was established in 1790, and detailed meteorological records, as well as records of sunspot observations, have been kept ever since. Deep in the vaults in the passageways underneath the observatory, Butler discovered a cache of these records going back to 1795. Plotting the variations in the sunspot cycle against variations in temperature, he found that the two sets of records matched incredibly well. For Butler, this not only confirmed the findings of Walter Maunder: it has relevance for the debates on climate change today. In the last few solar cycles, activity has been higher than normal – perhaps corresponding to the global warming also attributed to the increase of so-called greenhouse gases in the atmosphere. Butler claims that 'the changes in the Sun are one of the principal causes of the change in temperature of the Earth in recent decades'.

If Butler's assertion that solar activity plays a major role in Earth's weather and climate is true, then scientists will need to find out how this can be so. One mechanism by which it can come to bear on our weather is by affecting the clouds. The idea is that when a sun storm hits the Earth, it can electrically charge the tops of clouds, high in the atmosphere. Clouds are made of water droplets and ice crystals. When a cloud is electrified, electric currents pass through it and this can affect the relative populations of water droplets and ice crystals. Brian Tinsley and his colleagues at the University of Texas are carrying out research into the connection between solar activity and clouds; he explains: 'The electrical currents charge up the droplets at the tops of clouds where they are very cold, and a very small amount of electric charge has the capability to make them freeze.' The more ice crystals there are at the cloud tops, the greater is the likelihood of rain. And with rain, the cloud disappears, allowing the Sun to increase the rate at which it heats the Earth's surface. So the theory goes that sun storms arriving on Earth have the effect of reducing cloud cover, allowing the Earth to warm up.

To test that theory, Tinsley and his team take to the skies whenever auroral activity – and therefore solar activity – is high. They climb to the tops of clouds in an aeroplane that is specially equipped for carrying out high-altitude weather investigation. They sample the cloud directly, collecting water droplets and ice crystals using two metal tubes attached to the aeroplane's fuselage. Detectors in the tubes measure how much electric charge is carried by the cloud particles – to see whether the increased electrical activity leading to the auroras can have effects atmosphere-wide. So far, the results do seem to suggest that there is a greater concentration of ice crystals during periods of intense auroral activity – there may well be a link between sun storms and our weather. Weather is a day-to-day phenomenon, but climate is the average weather conditions over the longer term. Long-term variations in solar activity really could play an important part in climate variations. And solar physicists have confirmed that the actual power output of the Sun varies with the sunspot cycle – so not only do clouds clear, allowing the Sun to warm the Earth more efficiently, but the amount of energy reaching the Earth in the first place is greater, too.

Even today, astronomers are not sure exactly what causes sunspots, or quite how they relate to solar activity. But it is known that they are regions – each normally larger than the Earth – in which the gas is slightly cooler than in the surrounding areas: they are at a temperature of about 4,000 degrees C compared with 5,500 degrees C on the rest of the surface. This explains why they look dark against the rest of the Sun's surface – because they are cooler, they radiate less light. If you looked just at the sunspots, isolated from the rest of the surface, they would be extremely bright. It is also known that sunspots are intimately linked with the activity of the Sun's magnetic field. They form where magnetic field lines are vertical with respect to the Sun's surface – neighbouring sunspots have opposite magnetic polarity, because magnetic field lines are looping out of the surface and back in again. Loops of magnetic field that rise above the surface can form huge tubes of hot plasma that hang thousands of kilometres above the surface. These looped tubes of hot gas are called prominences.

The reason the Sun's magnetic field is so complex compared with that of the Earth has to do with the way the Sun rotates. On the Sun, some parts of the surface take longer to rotate than others. This is not true on Earth: because its surface is rigid, all parts of the world turn around once in twenty-four hours (actually, 23 hours, 56 minutes and 4.1 seconds). However, the situation on the Sun is quite different: regions close to the solar poles take about ten days more than those near the equator to complete each rotation: the period of rotation varies from twenty-five to thirty-five days. The reason for this difference is that, unlike the Earth, the Sun is made of gas. It seems, however, that the inside of the Sun behaves as a rigid ball because of its high density. As it spins, the rigid ball drags the upper, gaseous levels with it. And although the time it takes to rotate once is the same for the whole of this rigid ball, the actual speed at which the surface of the ball moves is quicker at the equator than near the poles. And so it drags the gaseous layer around faster there, too. The same differential rotation is observed on the planets Jupiter, Saturn, Uranus and Neptune, which are also made of gas.

Dr Nicola Fox explains how the differential rotation of the Sun can lead to its complex magnetic field: 'If you imagine twisting up a long piece of string round and round and round, you can see that the configuration is going to become very confused very quickly.' As the magnetic field lines become more twisted, they begin to loop out of the Sun's surface, and – so the theory goes – the result is sunspots, solar flares and a general increase in solar activity. One further oddity of the Sun's magnetism concerns its magnetic poles. While the Sun does generally have a magnetic north pole and a magnetic south pole, because of the complex nature of the magnetic field it sometimes has two north poles or two south poles.

Doing the splits

Astronomers use a wide variety of techniques to study the nature of the Sun's features such as sunspots, solar flares and coronal mass ejections. One of the most important is spectroscopy – the analysis of light by splitting it up into its component colours, to form a spectrum. This technique has its origins in the nineteenth century, although it depends upon a discovery made nearly 150 years previously, by Isaac Newton. In the 1660s Newton was investigating the rainbow-like spectrum produced when white light, such as light from the Sun, passes through a glass prism. Before Newton, scientists had reasoned that the spread of colours in a white light spectrum were somehow introduced by the prism itself. But in a classic investigation – which he referred to as his 'crucial experiment' – Newton used two prisms, recombining the spectrum of the first to produce a spot of white light. During part of this investigation, he blocked off all but the blue light of the spectrum produced by the first prism; in this case, there was neither a spectrum nor a spot of white light produced by the second prism. The colours that recombined to make white light in the first stage of the experiment were the colours present in white light in the first place. Newton realized that white light is actually a mixture of all the colours seen in the white-light spectrum, and that the prism simply split up the light by bending it to different degrees according to its colour. Contrary to popular belief, there are not just seven colours in the spectrum: there is a continuous range of colours, from red to blue.

In the early part of the nineteenth century, German physicist Josef Fraunhofer used prisms to investigate the spectrum of sunlight, and of the light from other stars and from the planet Venus. He noticed that the spectrum of light from these objects was crossed by a number of dark lines, where specific parts of the spectrum were missing. This fact had been discovered before, but Fraunhofer was able to show that the lines had fixed positions in the spectrum, and he catalogued a total of 574 of them. The colour of light depends upon its wavelength – you can fit about 2,500 wavelengths of blue light, but only 1,430 wavelengths of red light, into one millimetre, for example. And so Fraunhofer catalogued not only the 574 dark lines, but also the wavelengths of the missing light.

Fraunhofer and many other nineteenth-century physicists also investigated the spectra of light given out by hot substances, such as metal vapours. Because the light was actually given out by these hot materials, this type of spectrum is called an emission spectrum. These spectra consist of mostly darkness, with just a few bright lines. The bright lines are like a fingerprint of the chemical elements present – you can faithfully identify an element by its emission spectrum. It became clear that the dark lines in the Sun's spectrum corresponded to some of the bright lines observed in emission spectra. Experiments on Earth showed that when you shine white light through a hot gas, you get dark lines, just like those observed in the solar spectrum. The reason for this is that the gas absorbs certain colours – the same colours that it emits. It re-radiates the light that it absorbs, but in all directions, and this is why the spectrum appears darker at those points than across the rest of the spectrum. In fact, a spectrum like that produced from sunlight is called an absorption spectrum. In 1861, armed with this knowledge, German spectroscopists Gustav Kirchoff and Robert Bunsen were able to make the first guess of what elements were present on the surface of the Sun, 150 million kilometres away.

Spectroscopists were able to discover several chemical elements that were previously unknown. The greatest triumph in this endeavour began during the solar eclipse of 1868, when French astronomer Pierre Janssen analysed the emission spectrum of the Sun's corona – which is visible during a solar eclipse because the Moon obscures the Sun's disc. Janssen detected a previously unknown line in the spectrum. Several scientists became involved in the quest to identify the line, and to attribute it to a known element. However, it turned out that the line corresponded to an element that had not yet been discovered. It was named helium, after the Greek word for the Sun, 'helios'. The other main constituent of the Sun, hydrogen, produces several lines across the spectrum. The most important one is red, and is referred to by astronomers as 'hydrogen alpha', or 'Hα'. Between the Sun's surface and the corona is a layer that appears red because there is a great deal of Hα emitted there. This part of the Sun is called the chromosphere (coloured ball). The chromosphere is only about 10,000 kilometres thick, and it is visible, in profile, during a solar eclipse. Solar prominences, thrown up by the Sun's tortured magnetic field, extend from the chromosphere into the corona.

One of the most important developments in solar observation was the spectroheliograph, an instrument that revolutionized the study of the Sun. It was invented in 1889 by American astronomer George Ellory Hale, who was also the first person to discover the magnetic fields associated with sunspots. The spectroheliograph enabled astronomers to produce images of the Sun using particular wavelengths of light. This is the equivalent of putting a filter over a camera to allow only certain colours through. Indeed, astronomers are now able to use a wide range of very specific filters and other instruments that help produce images of objects from the light they emit in a very specific region of the spectrum. Of particular interest were pictures taken in Hα, which enabled astronomers for the first time to see the chromosphere over the disc of the Sun, not only at its edge during a solar eclipse. The chromosphere is alive with interesting features. Across the solar disc prominences appear as long thin trails called filaments, and in parts of the chromosphere above sunspots there are bright patches called plages. The whole chromosphere looks like a translucent ball of worn red velvet.

As well as producing specific, individual wavelengths of light – such as hydrogen's red Hα – hot objects like the

Sun emit light across a continuous range of wavelengths. This is called incandescence (mentioned earlier in connection with a candle flame), the process behind things that glow red hot or white hot. The element of an electric hob heats up to red hot, while the filament of a light bulb is white hot – both are incandescent. The range of wavelengths emitted by an object is characteristic of its temperature. The element of an electric toaster emits only red light and a little orange and yellow light – the low-energy end of the spectrum. An incandescent light bulb emits all the colours of the white light spectrum, including red, orange and yellow light, but also green and blue light. The white light spectrum contains all the colours visible to the human eye, but it is just a tiny part of a far wider spectrum: the electromagnetic spectrum. So light is one of a family of different types of electromagnetic radiation. Beyond the short-wavelength, blue end of the white light spectrum lies ultraviolet, then X-rays and finally gamma rays. Beyond the longer-wavelength, red end are infrared and radio waves. An object that is hotter than the filament of a light bulb emits all the colours of the visible spectrum but also ultraviolet; if it is really hot, it may even emit X-rays or gamma rays. The Sun's surface gives off visible light, but is not hot enough to produce much ultraviolet, and emits virtually no X-rays. The much hotter chromosphere is very strong in ultraviolet, and the corona, at 2 million degrees C, is strong in X-rays.

A wider view

During the twentieth century, astronomers learned how to extend their investigations of the Universe to include these other types of electromagnetic radiation. Radio waves were investigated first: huge parabolic dishes focus the radio waves from distant stars and galaxies on to a receiver. But now there are infrared, ultraviolet, X-ray and even gamma ray observatories, both here on Earth and in space. The main advantage of placing an observatory in space – that observations do not have to be made through the atmosphere – is particularly important for ultraviolet, X-ray and gamma ray wavelengths. Most ultraviolet wavelengths are absorbed strongly by oxygen and nitrogen – which between them make up nearly all of the atmosphere – and much of the rest is absorbed by ozone. Fortunately for animal and plant life, hardly any X-rays and gamma rays make it through the atmosphere. Nearly all radio waves penetrate the atmosphere as if it were not there and, of course, visible wavelengths make it through – otherwise we would not be able to see the Sun, the Moon, the planets and the stars. However, even for wavelengths that make it through the atmosphere relatively unimpeded, it often makes sense to observe from space. For extremely sensitive measurements in the radiowave part of the spectrum, for example, installing a telescope in space means that observations are made out of reach of the 'radio noise' caused by mobile telephones and television transmitters.

The most famous observatory in space is the Hubble Space Telescope, launched in 1990, which has brought a wealth of new astronomical discoveries. It carries instruments for producing images from visible light, infrared and ultraviolet. There is a spectrograph on board, to gather information about the spectrum of distant objects and to take pictures using any particular wavelength of light. Incredible photographs of the stars and planets taken using the spectrograph have brought dramatic new scientific insights, as well as providing the people of Earth with some stunning images of the Universe in which they live.

However, the Hubble Space Telescope does not look at the Sun – its specialities are planets and deep-space objects such as nebulas and galaxies. A number of space-based observatories have been designed to create a better understanding of the Sun, and some of the latest ones are bringing a chance to forecast space weather, perhaps giving us here on Earth more of a chance to prepare ourselves for a sun storm. Beginning in the 1960s, a series of spacecraft called OSOs (Orbiting Solar Observatories) took advantage of the lack of atmosphere in orbit, and gathered information about the Sun's X-ray, ultraviolet and gamma ray emissions. In 1971 the seventh OSO brought the first evidence of coronal mass ejections. X-ray telescopes on board the American space station Skylab were the first to examine the corona in detail in 1973. From the ground, the Sun's surface can be blocked out, by placing a disc of the right size in front of a telescope. But apart from fleeting moments during total solar eclipses, astronomers on the ground had always had to view the corona accompanied by the light of the sky. From Skylab's position in orbit above the atmosphere, it could study the corona against a black sky for the first time. The images produced on Skylab showed that there are huge 'holes' in the corona, which appear as dark patches on X-ray photographs. It has been found that the solar wind from coronal holes is twice as fast as that which comes from the rest of the Sun. The instruments aboard Skylab produced the first clear images of the Sun in the X-ray part of the electromagnetic spectrum – quite a task, since X-rays are absorbed by glass lenses and mirrors. Since 1978, with the launch of the Einstein Observatory, X-ray telescopes have used concentric metal cylinders to focus X-rays on to a detector by glancing reflections, and detailed X-ray images can now be obtained.

Also in 1978, ISEE3 (International Sun–Earth Explorer 3) was launched into space. This particular mission is notable because the spacecraft was flown into a special type of orbit, called a 'halo' orbit. Most space telescopes orbit around the Earth, which means that they regularly pass behind the Earth, losing sight of the Sun. One way to get a constant view of the Sun is to orbit not around the Earth but around the Sun itself, as the planets do. While this would indeed give the spacecraft a constant view of the Sun, it would mean that for months at a time contact with the spacecraft would be lost, as it passed behind the Sun as seen from Earth. This happens with all the planets, too. In a halo orbit, the gravitational forces exerted on a spacecraft by the Sun and by the Earth are equal, and this has the effect of keeping the spacecraft in view of both the Sun and the Earth at all times. This position relative to the Earth is called a Lagrangian point, after French mathematician Joseph-Louis Lagrange, who predicted their existence in 1772.

ISEE3 studied solar flares using an X-ray spectrometer. There were several other missions involving space-based solar observatories during the 1980s, including the SMM (Solar Maximum Mission), launched to coincide with the maximum of solar cycle number 21. The SMM spacecraft carried a range of instruments designed to study the Sun in ultraviolet and visible wavelengths.

Like the 1980s, the 1990s were an exciting time in space-based solar research. In 1990 an ambitious project called Ulysses sent a space probe into an orbit that takes it over the Sun's north and south poles, and extends out as far as Jupiter's orbit, 778 million kilometres from the Sun. Built by the European Space Agency and launched by NASA, the Ulysses spacecraft detects high-speed streams of energetic particles not detectable from Earth. In 1991 it was joined in its quest to discover more about the Sun–Earth system by a Japanese probe, Yohkoh (sunbeam), that carried a range of instruments for analysing the X-ray radiation coming from the Sun. Space-based research into the Sun was hotting up: in November 1994, as part of the International Solar-Terrestrial Program, NASA launched Wind. The instruments aboard Wind act like a space weather vane, and can warn other spacecraft of changes in the solar wind. Another important mission was Cluster, a quartet of satellites designed to study the interaction of the solar wind and coronal mass ejections with the Earth's magnetic field.

Group effort

The idea behind Cluster was to use the data from four satellites together, to create a three-dimensional picture of how sun storms affect the magnetic environment around Earth. The project involved 2,000 scientists from around the world, took ten years of planning and cost £200 million. The satellites were launched aboard an Ariane-5 rocket, from Kourou in French Guiana, South America, in June 1996. Mike Lockwood of the Rutherford Appleton Laboratory says the launch caused great excitement among scientists: 'The whole community, globally, was glued to a screen of some kind to see the launch.' NASA's Nicola Fox was also involved in the Cluster programme: 'As the rocket lifted up, it was just a whole bag of mixed emotions: there was relief that it was finally on the rocket and going up, there was excitement about the type of science it was going to show us.'

The hopes of the scientists who had worked on Cluster were dashed when, just thirty seconds after lift-off, the rocket – and the satellites it was carrying – exploded 3.5 kilometres above the ground, travelling at nearly 860 kilometres per hour. The resulting pieces were scattered over a wide area of the Guianan swamps. Lockwood remembers the effect it had on the people whose work had just been obliterated: 'People were crying around the room, and there was a touch of anger – this should not have happened to the community of scientists who were so looking forward to the data coming in.' To make things worse, the satellites were not insured. The enquiry into the incident found that the problem was with the guidance system, which was using computer software designed for use with an Ariane-4 rocket.

In 1997 there was good news for the scientists involved in Cluster: the satellites are to have a second chance. Using some spare parts from the first mission, the project is going ahead, scheduled for launch in the summer of 2000.

Meanwhile, another spacecraft designed to study the interaction between the Sun and the Earth's magnetic field was launched, without a hitch, in December 1995. Called SOHO (Solar and Heliospheric Observatory), it travels around the Sun in a halo orbit like the ISEE3 spacecraft described above. Aboard SOHO are three sets of instruments, each with a specific task. One set of instruments takes measurements of the composition and speed of the solar wind that blows past it. Analysis of the solar wind, which originates in the corona, should reveal a great deal about how the corona can be so much hotter than the surface beneath it – one of the great mysteries of solar science. The puzzle concerns the fact that the boundary between the surface and the corona seems to disobey a scientific law: if two things are in contact, heat passes from hot to cold, and yet the relatively cool surface is heating the searingly hot corona. This has been pondered over since scientists first discovered the temperature of the corona in the 1940s. There must be some mechanism by which energy is reaching the corona from below the surface, and SOHO may be the one to find out what it is.

Another set of instruments on board SOHO investigates the nature of the corona more directly. One of the instruments is the Extreme Ultraviolet Imaging Telescope, which produces images of the Sun at wavelengths within the ultraviolet range of the spectrum. The telescope acts as a spectroheliograph, because it looks at specific wavelengths produced by the Sun. The wavelengths chosen correspond to particular elements within the corona, and the resulting pictures are awe-inspiring: they highlight the corona, but leave the surface almost totally black, giving these solar portraits a ghostly look. The corona is highlighted in a different way by another type of instrument – a coronagraph, a telescope that has a disc in front of it to block out the Sun's surface. The final set of instruments looks at the Sun's surface itself. They watch the oscillations of the Sun's surface when violent events take place there. It has been known for a long time that the Sun 'rings' like a bell after any kind of disturbance, sending shock waves in all directions across the solar surface and through its centre. By analysing the oscillations, solar physicists should be able to deduce a great deal about the nature of the Sun beneath its surface. This is like seismology, which uses analysis of the oscillations of the Earth caused by earthquakes to investigate the Earth's mantle and core. And so this method of peering beneath the surface of the Sun is called helioseismology.

Another exciting venture using helioseismology to probe the inner workings of the Sun is the GONG project – the Global Oscillation Network Group. This international effort employs six identical helioseismological stations – in California, Hawaii, Australia, India, the Canary Islands and Chile. The fact that they are spread across the globe means that they can provide constant measurements of the Sun – when it is night in one of the centres one or more of the others will then be in daylight. The instruments are designed to detect a particular oscillation of the Sun's surface – one that takes five minutes to complete. The five-minute oscillation promises to reveal new details about the solar interior.

Results obtained by SOHO during 1996 showed dramatic activity on the Sun, producing exciting pictures of coronal mass ejections, for example. Animations made from sets of time-ordered photographs showed the Sun spewing out matter from the chromosphere and the corona. This matter was seen being thrown out at the sides of the solar disc as seen from Earth, so there was no cause for concern, as Mike Lockwood explains: 'Although they're very spectacular to watch, in terms of worrying about the disturbances here on Earth they're not a problem because they're going to miss the Earth.'

SOHO and the other probes were joined in space by another spacecraft in February 1996: Polar. Rather than gazing sunwards as the others do, Polar looks at the Earth's poles, to keep a track of the auroras and disturbances in the Earth's magnetic field.

With all the spacecraft in place, on 7 April 1997, the Sun flared up again. This time, the appearance of the eruption was different. Instead of being directed out in a particular direction, or just at the sides, the Sun blew material off all around, like a halo. This meant that the solar matter was blown off in the direction of Earth, as Nicola Fox explains: 'You could see the circle getting bigger and bigger and this was the sun storm coming towards us.' One of the instruments on board SOHO showed how this violent event made the Sun reverberate, with a powerful shock wave that originated on the side of the Sun facing Earth. This was more evidence that the coronal mass ejection was directed towards Earth. Within three or four days, the matter would reach our planet.

There's a storm brewing

As a result of this observation, the scientific community was on alert, fearing huge magnetic storms on Earth, like those that ravaged Quebec in 1989. The SEC (Space Environment Center) decided that the April 1997 solar event would have no ill effects, and issued no warnings. But the US Air Force was taking no such chances, and it warned military units to expect interruptions in their communications. The reason why no definite forecast could be made is that not all ejections from the Sun have harmful effects on Earth. This is partly because the Earth is such a small target in the vast open space of the Solar System; but even when they do reach Earth, the bubbles of plasma are sometimes rebuffed by the Earth's magnetism, passing harmlessly by. The determining factor is the polarity of the plasma cloud: if it is the same as the direction of the Earth's magnetic field, then the cloud is repelled. If the plasma cloud's magnetic field is polarized in the opposite direction from Earth's magnetic field, the plasma cloud can approach much closer to the Earth, and that is when it can create havoc at ground level.

Despite the SEC's decision not to issue warnings, NASA scientists contacted the media, which broadcast news stories about the sun storm. The telephones of the Space Environment Center began to ring, bringing a stream of questions from individuals and organizations worried about the potential effects. Ernie Hildner of the SEC remembers: 'We had a hospital administrator call up and say, "Of course, we will turn off all our computers, including the medical monitoring computers at the time the storm is supposed to hit" ... Kennedy airport people called up and said, "Should we keep 747s on the ground?"' The telephones were ringing. at NASA, too, as Nicola Fox remembers: 'We had things like, "Will my plants die?" "Should I switch off my computer for the next two days?" "Should I bring my pets inside?"' Everyone made ready, and waited for the storm to disrupt their lives. But this time, the solar plasma swept past the Earth without incident.

In August 1997 NASA launched the Advanced Composition Explorer (ACE). This spacecraft carries a range of instruments, including two magnetometers that determine the polarity of approaching plasma clouds. According to Ernie Hildner, data from ACE would enable the Space Environment Center to make predictions with almost a hundred per cent accuracy one hour in advance. He says that 'customers who are affected by geomagnetic storms will very much appreciate having one hour's notice'. And, since the launch of ACE, the accuracy and sophistication of space weather predictions have increased. In 1999 the Space Environment Center's parent organization, the National Oceanic and Atmospheric Administration, issued a set of 'space weather scales', akin to the Richter scale for earthquakes or the Beaufort scale for wind speed. There are three different scales: one for geomagnetic storms, one for solar radiation storms and one for radio blackouts. Each scale runs from one to five. So, for example, a G5 geomagnetic storm can cause electricity grid systems to collapse and cause problems in tracking satellites and the auroras to be seen as far from the poles as the equator; while a G1 causes mere fluctuations in electrical power supplies and can affect the migration of birds, which depends upon the Earth's magnetic field.

The unprecedented views of the Sun that SOHO, ACE and the other probes supply are beginning to form a clearer picture of how the twists and turns of the Sun's magnetic field can produce the variation in solar output, and how that variation can affect us here on Earth. Having studied the Sun's activity for a number of years, they have provided evidence that the Sun is very active at all times, even during the minimum of the solar cycle. And in 1999 they showed that even during the height of the solar cycle, the Sun can have quiet days. On 12 May, the Wind and ACE spacecraft measured the density of the solar wind as only 2 per cent of its normal value. Without the usual buffeting from the solar wind, the Earth's magnetosphere could expand to five or six times its normal size.

In 1996 scientists based at the Rutherford Appleton Laboratory, who are part of the SOHO programme, discovered a previously unknown feature on the surface of the Sun. During a time when the Sun was particularly 'quiet', they observed flashes of activity that they called 'blinkers', within the granulations created by convection at the Sun's surface. The scientists estimated that about 3,000 blinkers occur at any time across the whole surface of the Sun, each one about the size of Earth and releasing energy equivalent to the explosion of about a hundred million tonnes of TNT. The team believes that blinkers could be the mechanism behind that mystery of solar physics, the heating of the corona. Shortly before the British team discovered blinkers, a team at the Stanford-Lockheed Institute for Space Research in Palo Alto, California, had discovered an effect that they refer to as a 'magnetic carpet'. The scientists describe how relatively small but very numerous loops of magnetic field protrude above the Sun's surface, each bringing as much energy as a large power station would generate in a million years. It is because blinkers and the magnetic carpet occur across the whole solar surface that they are creating excitement among the people studying the Sun. The solar wind spreads out in all directions, throwing particles out at speeds higher than the 500 kilometres per second necessary to escape the Sun's gravity. And to do this, it needs the kind of energy supplied by the extremely high temperatures found in the corona.

With our understanding of the Sun advancing so rapidly, and with ACE and the other solar probes watching the Sun from space, the future of more accurate space weather forecasting looks assured. Perhaps within the next few years, the state of the space environment will be included in our television, radio and newspaper weather bulletins. But according to Ernie Hildner, ACE might be vulnerable to the very thing it is measuring: 'The situation with ACE and sun storms is very like the situation for an anemometer – a wind velocity meter – trying to measure a hurricane: it can be ripped off its moorings, and taken away. Unfortunately, ACE out there helping us measure sun storms about to hit Earth could be damaged by just one of those storms.'

And whether we can accurately predict space weather or not, violent sun storms can disrupt our increasingly technology-dependent way of life. As Professor Mike Lockwood says, 'We could actually find ourselves unable to do things that we've been able to do for most of the past century, because we no longer have the necessary tools to do it; that these things have been removed from us in a stroke.' Many navigation techniques on Earth depend upon a GPS (global positioning system) of satellites. Communications between these satellites and instruments on the ground could be compromised by severe geomagnetic storms. And Dr Daniel

Baker reminds us that sun storms can threaten military communications and satellite tracking: 'There could be a rather catastrophic effect on much of civilization as we know it, if one of these storms occurred at the wrong time in the wrong place.' Perhaps this indicates that we have come to depend on technology in ways that we might regret.

BLACK HOLES
... searching for nature's
ultimate abyss...

Until recently, black holes were hypothetical objects – confined largely to theoretical physics and science-fiction stories. They were a by-product of our increasing understanding of the Universe, a mathematical inevitability, which yet defies our common-sense notions about the nature of space and time. As we shall see in this chapter, a black hole is very strange indeed. But, despite some scepticism, astronomers now in the scientific community, have compelling evidence that these weird objects really do exist in our Universe.

A star is born again

In 1992 a little-known star called V404 Cygni shot to fame in scientific circles when astronomers claimed that it almost certainly was a black hole. Three years earlier a Japanese orbiting observatory, Ginga, had detected a powerful surge of X-rays coming from the star, in the very distant constellation of Cygnus. Along with the X-rays, the star was emitting more visible light than normal, too. It was still far too faint to be detected by the unaided eye, despite growing more than 600 times brighter. Calculations based on the radiation received from the star led to the almost inescapable conclusion that V404 Cygni consisted of a giant star orbiting a black hole. The burst of radiation observed in

1989 – and also, for the first time, in 1938 – was caused by gas heating up as it was pulled off the giant star by the intense gravitational forces of the black hole.

Gravity is the key to understanding black holes. It is the mechanism behind the making of these strange objects, and is also responsible for their behaviour. And it was as an inevitable consequence of the laws of gravity that the concept of black holes originated in the first place. Once you understand that everything in the Universe is affected by gravity and you know how gravity makes things behave, the concept of black holes jumps out at you, begging to be noticed. Even so, the predicted behaviour of these objects is so much at odds with the rest of the physical world that no one could confidently assume that they exist. Since the 1970s more and more evidence has accumulated in favour of the existence of black holes, and astrophysicists now generally accept that they do exist. A number of probable candidates have been discovered, including V404 Cygni.

Before V404 Cygni became a black hole, it was an ordinary massive star – with a mass several times greater than the mass of our Sun. Like the Sun, it formed thousands of millions of years ago from a cloud of interstellar gas and dust – a nebula – brought together by gravity. The force of gravity between any two objects pulls them closer together. When there is a large number of objects – the particles of gas and dust in a nebula, for example – this has the result of bringing everything together towards a common point in space. The Sun began life in this way, as gas and dust from a nebula clumped together in space about 5,000 million years ago. Gravity is tenacious: the force between two objects increases as they move closer together, so, once the gravitational collapse of the solar nebula started, it accelerated, and the clump of matter became more and more dense. This 'protostar' heated up, especially at the centre, where the gases were crushed inexorably. As explained in 'Sun Storm', once the temperature rose sufficiently at the core of this protostar, nuclear fusion was initiated, producing huge amounts of heat. This heat tended to expand the core, and this halted the gravitational collapse. The Sun is in its 'main sequence' – the phase of its life cycle in which the gravitational collapse is offset by the reactions at its core. Eventually, the nuclear reactions will cease, and gravity will take over once again.

As the Sun begins to shrink at the end of its main sequence, its core temperature will increase and, since there is no more hydrogen to fuse into helium, the heat will pass to the Sun's outer layers. Hydrogen in these outer layers will begin to fuse into helium, and this in turn will heat up the centre, where a new set of nuclear fusion reactions may take place. The helium in the core will undergo fusion into heavier elements – carbon first, and then oxygen. At this stage, another force will come into play to fight against the gravitational collapse: electron degeneracy. It was discovered in the 1920s that certain particles, including electrons, resist being squashed into a small space as a result of strange rules that govern their behaviour. The force that results is known as electron degeneracy pressure, and this will prevent any further gravitational collapse, leaving a core in which no more fusion can take place, and that will shrink no more.

And so the Sun will end up as the type of dead star called a white dwarf, composed of carbon and oxygen nuclei, that slowly cools as it sits redundant in space, held up against gravity by electron degeneracy pressure in plasma. Meanwhile, the outer layers, in which there will still be some fusion taking place, will be thrown off into space – first swelling and then ending up as a distended cloud of gas around the white dwarf. In the case of the Sun, the expanding gases will extend beyond the orbit of Venus, and possibly out as far as Earth's orbit. The white dwarf remnant at the centre will have a diameter about equal to the diameter of the Earth and a density about 50,000 times that of gold.

Many stars exist in pairs or groups, held together by their mutual gravity: they are in orbit around one another. If a white dwarf is a member of a pair – or 'binary system' – gravity pulls gas towards it from the outer layers of its partner star. When the gases arrive at the surface of the white dwarf, they may take part in nuclear fusion – the white dwarf temporarily comes alive again – and this results in a sudden brightening of the star as seen from Earth. If the binary system was not visible to the unaided eye from Earth before the eruption, but becomes so afterwards, it will appear as a new star in the sky. For this reason, such an occurrence was given the name 'nova', from the Greek for 'new'. The star called V404 Cygni is a kind of nova, but there is a black hole – and not a white dwarf – in this binary system. Astronomers know that it cannot be a white dwarf, because it is too massive. A star that is much heavier than the Sun cannot become a white dwarf, because due to the greater mass the gravitational collapse can overcome even the electron degeneracy pressure.

When massive stars come to the end of their main sequence, they fuse helium into carbon and oxygen as described above, but fusion does not stop there. The star is hot enough to build up heavier elements, including silicon, sulphur and calcium. This continues until the element iron is produced. When iron fuses into heavier elements, it does not release huge amounts of energy as the previous phases of fusion do. In fact, the iron takes in huge amounts of energy, the star cools and it can no longer be supported against gravity. The outer layers collapse, then are driven out into space accompanied by a huge burst of radiation – which may be visible from Earth as a 'supernova'. Seven such supernovas have been observed in recorded history, the most famous being one that occurred in 1054: it was noted by astronomers in China and Korea and appears on rock paintings in the Americas. This supernova was bright enough to be seen during the day.

Once a star has 'gone supernova', its core may collapse still further, to become a 'neutron star'. Neutrons are found in ordinary matter: together with protons, they make up the atomic nucleus around which electrons orbit. At the centre of a massive collapsing star, however, particles of these three types are crushed together so hard by the gravitational collapse that the positively charged protons and the negatively charged electrons combine, forming more neutrons (which have no electric charge). So, apart from an outer shell of solid iron and an inner core of exotic subatomic particles, the star has become composed almost exclusively of neutrons – hence the name. A neutron star may have a radius of about twenty kilometres and yet have a mass equal to several Suns – its density is a million million times that of gold. A pinhead-sized piece of the material of a neutron star on Earth would weigh as much as a supertanker.

Most neutron stars spin rapidly – a remnant of the rotation of the original star, but at an increased rate due to the fact that the star has shrunk. (This is similar to what happens when spinning ice skaters bring their arms in close to their body to increase their rate of spin.) A neutron star has an intense magnetic field, and this causes protons and electrons around the surface of the star to emit radio waves.

The radio waves form a powerful beam emanating from the neutron star's magnetic poles. As the star spins, these beams sweep across space like a cosmic lighthouse. From Earth, this can be detected as a pulsing source of radio waves – a pulsar.

Most neutron stars collapse no further because of 'neutron degeneracy', which is like the electron degeneracy that halts the collapse of white dwarf stars, only stronger. And yet if the mass of the neutron star is large enough – above about three times the mass of the Sun – even neutron degeneracy will be overcome, and the gravitational collapse will continue unhindered. The result will be an object that keeps on collapsing, with nothing to prevent it crushing itself out of existence. In fact, the object's mass becomes concentrated in ever smaller volumes, and its density increases dramatically. Ultimately, it will shrink to a point of no size – and infinite density. Such a hypothetical point is called a singularity, and its predicted existence is one of the great problems of physics. The gravitational field near a singularity is so strong that even light cannot escape – this is a black hole.

No escape

As explained in 'Day Return to Space', a rocket fired from Earth that attains a speed of 11 kilometres per second has enough energy to escape from the planet's gravitational influences. Any slower, and the rocket will have to go into orbit around Earth or else it will eventually fall back down to the ground. Light is affected by gravity, just like everything else. If you shine a torch straight up into the air on a clear night, the light that makes it through the atmosphere (without hitting air molecules or dust particles) easily escapes the Earth, because it travels at 300,000 kilometres per second – much faster than the escape velocity. If you could shine a torch upwards from the surface of a neutron star, the light would actually be impeded by the intense gravitational field there. It would not be slowed down, however, because the speed of light is the one thing in the Universe that is absolutely constant. This fact was determined from the theory and experiment towards the end of the nineteenth century. So a light beam from a torch travelling towards you or away from you at any speed will still arrive at the speed of light.

Even so, a light beam from a torch at the surface of a neutron star would lose energy as it escapes from its gravitational field. This would manifest itself as a change in the light's wavelength. The wavelength of a light wave would increase as if the light waves were being drawn out. So short-wavelength blue light would be shifted towards the longer-wavelength red end of the spectrum, for example. This effect – called gravitational red shift – has actually been detected in the light coming from extremely dense objects far out in the Universe. What makes this happen is very strange: in a sense, the light really is stretched out, because the gravity near the neutron star actually stretches the space through which the light travels. Put another way, gravity stretches time, so that the time between oscillations of the wave is more than before, again making its wavelength longer. In fact, gravity warps 'space-time' – an inseparable combination of space and time that physicists use to understand the Universe.

The idea of space-time originated with German physicist Hermann Minkowski, in 1908, as an interpretation of Albert Einstein's 1905 Special Theory of Relativity. Before the principles of Einstein's relativity were worked out, physicists imagined that for the sake of equations governing an object's motion, the Universe worked something like a three-dimensional grid. Using Isaac Newton's laws of motion, they could work out how an object would move around this three-dimensional grid, according to how fast and in what direction it starts out and the forces that act upon it. This picture of the Universe assumed that 'now' is a meaningful concept across the entire grid, and that lengths and times are 'absolute' quantities that would be the same for any two people even if they were travelling relative to each other. In this Newtonian Universe, two beacons one kilometre apart and flashing one second apart will be measured as one second and one kilometre apart by anyone who sees them. Common sense tells us that this is what would happen. However, when it was found that the speed of light is constant, relative to any observer, this picture was under threat. In the 1890s Dutch physicist Hendrik Lorentz and Irish physicist George Fitzgerald were separately trying to reconcile the Newtonian Universe with the new discovery about the speed of light. As an attempt to hold on to the three-dimensional picture of things, they proposed that time ran differently for people moving relative to each other, and that lengths were different, too.

This idea was taken on by Einstein in his special relativity theory. According to this, different people travelling at different speeds relative to the two beacons above would see the beacons at different separations – both of time and of space. For one person, the beacons might be in the same place but flash at very different times, while the other sees them flash at the same time but separated by hundreds of kilometres. This is not an illusion – time and space really are different for people moving relative to each other. The separation between the two flashes is constant only when measured in four-dimensional space-time. The concept of space-time is difficult to comprehend, not least because it is impossible to imagine since our senses are restricted to the three-dimensional world that we perceive. However, it is a well-tested idea, and it is a useful way of explaining some of the scientific theory behind black holes, as we shall see.

The concept of space-time was particularly useful when it came to visualizing Einstein's General Theory of Relativity, published in 1916. In this picture, gravity is seen as a warping of space-time rather than a force in the conventional sense. Isolating any one, two or three dimensions makes it possible to picture the effect of gravity. So, for example, imagining three-dimensional space as a flat (two-dimensional) surface, gravity can be thought of as distorting the shape of the surface. This is where the popular image of space-time as a rubber sheet comes from (and we return to this image later in the chapter).

The General Theory of Relativity was a completely different way of interpreting the Universe from Isaac Newton's Universal Theory of Gravitation. Both theories predict the same results for most familiar situations – if Newton's theory was consistently wrong in 'everyday' use, it would not have survived for more than 200 years. Indeed, Newton's theory is still routinely used in many situations, as calculations based on it are simpler and quicker to perform than those of general relativity.

Still, the two theories were very different: Newton's idea was that gravity is a force between any two objects with mass – the force acts across large distances instantaneously. Einstein had to address the fact that his special relativity theory had shown that nothing – not even the transmission of gravitational force between two objects – can travel instantaneously. Everything takes a time to get somewhere else, and the Universe has a speed limit: the speed of light.

All forms of electromagnetic radiation travel at the speed of light – including radio waves, ultraviolet and X-rays. The result of Einstein's incorporating gravity into his relativity theory led to the idea that, rather than being a force, gravity is caused by the distortion of space-time.

To a physicist, the General Theory of Relativity 'feels' right because it is so mathematically elegant. But for such a radical ideological shift to become accepted as a central theory of astrophysics – and of science in general – Einstein's view of things had to prove itself. It has done so amply, by matching prediction to observation in many different situations in which predictions based on Newton's theory failed. One of these situations involves the deflection of light from distant stars as it passes close to the Sun. According to Einstein, light follows the curvature of space-time, and so is deflected as it travels past a massive object like the Sun. This should have the result that the stars from which that light has come appear in a slightly different location in the sky. The effect is very small with an object with the mass of the Sun, but would be much more marked for something like a neutron star. And light can actually be deflected completely around by a black hole, so that it ends up travelling back the other way, or even orbiting the hole a few times before either being sucked in or moving off into space.

The first person to suggest that light might be affected by gravity, in the same way as ordinary matter is, was British clergyman and scientist John Michell in the 1780s. Incidentally, Michell was the first to suggest the existence of objects possessing such a strong gravitational field that not even light could escape. The deflection of light by the Sun's gravity can be observed only by looking at stars very close to the Sun in the sky – those whose light passes very close to the Sun. It is normally impossible to see such stars, because of the daylight and the intense brilliance of the Sun. However, during a total solar eclipse the sky around the Sun is dark, and this makes it possible to see them.

An eclipse in 1919 gave supporters of general relativity a chance to test the new theory. The British physicist Arthur Eddington organized an expedition to Príncipe Island in East Africa, to record accurately the positions of stars close to the Sun in the sky during the eclipse. Comparing the positions of those stars when their light does not pass so close to the Sun – six months earlier, when the same stars were in the night sky – would allow any deflection to be measured. There were two predictions that could have come from Newton's theories. One would predict that light would not be deflected at all as it passes near the Sun, because light has no mass; the other would predict that there would be deflection, but only half as much as that predicted by Einstein's theory. (It is not necessary to go into the details of the mathematics here.)

The 1919 results gave confirmation to Einstein's theory and have been repeated, with more accurate instruments, many times since then. Many other strange predictions of general relativity have been confirmed, too, and the theory that the Universe is made of space-time that is curved around massive objects quickly became accepted. One is the idea that the bending of light around very massive objects, such as huge galaxies or black holes, can become so extreme that it would severely distort the images of more distant objects whose light passes close by them. This effect – an extension of the deviation of light passing close to the Sun – is called 'gravitational lensing'. In the 1970s the first example of a gravitational lens was discovered, and in 1999 ten classic examples of the phenomenon were found by searching through some of the images produced by the Hubble Space Telescope. The existence of black holes is another of the predictions of general relativity, and one of the reasons why people are so keen to find them is to find the ultimate verification of the theory.

When space is pictured as a rubber sheet – using the image mentioned earlier – you can imagine a heavy ball making a circular 'well' around it as it pulls the rubber downwards. Other, smaller, balls placed on the rubber sheet passing near to the heavy ball will move towards the large ball, as if they are being attracted to it. In the same way, according to general relativity, the presence of the Sun in space distorts space-time so that the Earth appears to be attracted towards it. The heavier the ball, the deeper the well and the steeper its sides. A neutron star distorts space so much that only very fast-moving objects can escape from this well. But the gravitational well caused by a neutron star does have a bottom – objects, and light, can emerge from the well if they are travelling fast enough. A black hole, however, forms a bottomless well. You can see why this is troubling to some people: once an object – or even light – travels into a bottomless well, it can never come out again.

Far from a black hole, an object can escape the intense gravitational field. The escape velocity increases as you move closer to a black hole, and the distance from the centre of a black hole at which the escape velocity is exactly equal to the speed of light is called the Schwarzschild radius. It is named after the German astronomer Karl Schwarzschild, who was the first to work out the shape of the space-time curvature around a massive object, using general relativity, in 1916. It was also he who, in the same year, predicted the existence and behaviour of black holes as a consequence of Einstein's theory. From anywhere within the Schwarzschild radius, nothing can ever escape. All the points around the centre of the black hole that are at this distance form the 'event horizon', so-called because no events that happen inside can ever be observed outside.

A black hole defies easy description. Professor Kip Thorne of the California Institute of Technology is one of the pioneers of recent black hole history. He describes it as 'not made of ordinary matter, not made of anti-matter, it's not made of matter at all: it's made of a pure warpage of space and time'. Professor Philip Charles, who was one of those who worked out that V404 Cygni is probably a black hole, says that 'nothing comes out of it, things only go into it; it's that one-way nature that probably upsets people most, causes them to be really disturbed'. And Professor Martin Rees, Astronomer Royal, based at the University of Cambridge, comments on the importance of the singularity: 'In the centre there is a so-called singularity, which is a place where … everything goes infinite; where, if you do calculations, as it were, the smoke pours out of the computer and something goes wrong. What that means is that deep inside black holes we have a signal that we need some fundamentally new physics to understand what is going on.'

Worrying though the concept of black holes is to some astrophysicists, their existence is now widely accepted. But, what would it be like near to a black hole?

Visiting the death star

Space probes have visited all but one of the planets of our Solar System. One day, a probe will no doubt be sent to the nearest star – a journey that will take tens or perhaps hundreds of years. And in the very distant future, we may visit a black hole – with extreme caution – but for now, this has to be a mission of our imaginations. From our safe vantage point some way from the event horizon, in an extremely rapid orbit around the black hole, we could send a probe down towards – and beyond – the event horizon. As the probe drew nearer to the black hole, it would be violently stretched by the gravitational field. This stretching is a result of what physicists call 'tidal forces'. The closer you are to something, the stronger the gravitational pull. So in ordinary Earth gravity your feet are pulled downwards slightly more strongly than your head, for example, and the difference between the two forces results in a very slight stretching of your body.

The gravitational field around a black hole is vastly more intense than that of the Earth. So the end of the space probe closer to the black hole will be pulled considerably more than the other end. It would certainly not be advisable for a person to venture too close: 'You or I falling into a black hole will get stretched, and our bodies will get mutilated before we even reach the horizon,' explains Kip Thorne. Astrophysicists have coined a memorable term for this effect, as Martin Rees says: 'As you get closer to the centre gravity will get more extreme, you will get squeezed and distorted – spaghettification is the name that some people use for this experience.'

The gravitational field would cause another interesting effect as the probe neared the black hole. The light struggling to escape from the surface of a neutron star is red-shifted because time runs more slowly in an intense gravitational field. And time would slow down for the probe, too – relative to the mother ship. It would appear to move more and more slowly from the mother ship's perspective. At the event horizon, time would be slowed down to a stop: the probe would hang at that location for ever relative to the mother ship. And yet, Kip Thorne tells us, 'If you are inside that spaceship looking back at me, you can wave, you can see me, you may think I can see you, but I can't; you go inside the event horizon you can still see out, you can still see me on the outside, right up to the point when you die.' You would still receive light signals coming in – they have no trouble going in that direction. Nobody knows what happens to objects that fall into a black hole. Once past the event horizon, a probe – if it was still in one piece – would be able to send back no information anyway.

Inside the event horizon, the probe would no doubt be pulled in the direction of the centre – towards the singularity, if such a thing exists. It was Karl Schwarzschild who first predicted the existence of the singularity in 1916. For the next few decades, as physics grew in scope and sophistication, it seemed increasingly likely that black holes did exist and had the kind of properties described above. Until the discovery of the first pulsar in 1967, however, black holes remained a theoretical speculation. As explained earlier, a pulsar is a spinning neutron star – not far removed conceptually from a black hole – and this is why the discovery of pulsars was so important. Many astrophysicists who were unable to accept the strange effects associated with black holes had convinced themselves that a supernova would always throw enough matter into space to prevent ever leaving enough matter to form a black hole or even a neutron star. These things were just too strange to comprehend.

It was British astronomer Jocelyn Bell who was the first to detect signals from a pulsar. She was using a radio telescope to search for other unusual objects called quasars. Studying the hundreds of metres of paper output from the radio telescope, she noticed an unidentified signal: pulses of radio waves with an incredibly regular and very rapid pulsing. After various theories about its origin were dismissed,

Bell wrote the letters 'LGM' next to the trace, to signify the possibility that the signals might be coming from a civilization of 'little green men'. When Bell and her colleagues worked out that this signal could only have been coming from a rapidly rotating neutron star, it was confirmation of the theory of the late stages of the evolution of a star. Bell's discovery of a pulsar confirmed the fact that neutron stars exist, and several more have been discovered; so why not black holes?

It was in the same year as Jocelyn Bell discovered the first signals from a pulsar that the term 'black hole' was coined – by American astrophysicist John Wheeler. But already in 1964 the first serious search for these objects had been undertaken, by Russian theoretical physicist Yakov Zel'dovich. The method used by Zel'dovich has much in common with the approach used by today's black hole hunters. He and his student Oktay Guseinov at the Institute of Applied Mathematics in Moscow scoured the star catalogues looking for binary stars. Their reasoning was straightforward enough: if one of the two stars in a binary system is not radiating much light, and has a mass greater than about three times the mass of the Sun, it is probably a black hole. This conclusion is a result of general relativity and other physical theories applied to super-dense objects: if the invisible partner has a mass less than three times the mass of the Sun, then the object is either a neutron star or a white dwarf. Astronomers had already developed ways to measure the masses of the two partners of a binary system and, from the catalogues, Zel'dovich and Guseinov highlighted five black hole 'candidates'.

In 1966 Zel'dovich, together with another Russian astronomer, Igor Novikov, came up with a giveaway sign that an object might be a black hole or a neutron star. They surmised that a massive, compact object such as a neutron star or black hole would pull huge amounts of gas off a partner star, and that this gas would heat up to enormous temperatures as it fell towards the massive object. At these temperatures, the gas would emit powerful X-rays, which we should be able to detect from Earth. And so Zel'dovich and Novikov suggested looking for binary systems with one partner emitting X-rays and the other emitting visible light. Using X-ray telescopes and optical telescopes together would be a good way of finding black holes or neutron stars in binary systems with ordinary stars. Calculating the masses of the two stars would then confirm whether one of them was a black hole.

In a binary system, stars rotate around a common centre – like two people holding hands, engaged in a whirling dance. If the two people in such a dance weigh about the same, they will both go round at the same speed, about a centre of rotation halfway between them. If one of the people is very heavy, and the other is much lighter, the heavy person will hardly move, while the lighter person will fly around at high speed. In this case, the centre of rotation is much closer to the heavy person. In fact, if you could not see the people, but you could somehow observe their motions, you would be able to estimate their relative masses. In the same way, astronomers can work out the masses of the stars in a binary system by working out how they move around each other.

In 1969 Donald Lynden-Bell, at the University of Cambridge, developed the idea of X-ray emission from black holes or neutron stars suggested by Zel'dovich and Novikov three years earlier. Assuming that gas leaks from the visible star in a binary system, as explained above, Lynden-Bell worked out what might happen as the gas approaches the black hole or neutron star. He found that the gravitational and magnetic fields around a neutron star or black hole would draw the gas into a specific configuration: a rapidly spinning 'accretion disc'. The temperatures near the centre of the disc may be as high as 100 million degrees C – hot enough to produce X-rays, as Zel'dovich and Novikov had proposed. However, in the 1960s, X-ray telescopes were in their infancy. To complicate this further, X-rays do not penetrate the Earth's atmosphere, so X-ray observatories would have to be sent to high altitudes – ideally into space. The first devices to detect extraterrestrial X-ray sources were carried aloft on high-altitude balloons and rockets. They were crude by today's standards, but they did show that X-rays were coming from space, and in 1970 the first true X-ray telescope – called Uhuru – was launched into orbit by NASA.

Swanning around

To illustrate the lengths to which astronomers are prepared to go in search of black holes, consider the story of perhaps the most famous black hole candidate, Cygnus X-1. Uhuru detected a source of X-rays – the first one to be discovered in the constellation of Cygnus (the Swan) – during a systematic survey of the sky. The X-ray source seemed to be very close to a known star, about 6,000 light years distant, with the less than memorable name of HDE226868. Astronomers suspected that there might be a star and a neutron star or black hole in orbit around one another, so they set about determining whether HDE226868 was in orbit around something. They did this by analysing the light coming from it. If the star is orbiting something, and the orbit is side-on or nearly side-on, then it will be moving towards Earth in one part of its orbit and then away from Earth in another part. When the motion is towards the Earth, the light from the star will be blue-shifted, and when the star is moving in the part of its orbit that takes it away from the Earth, its light will be red-shifted. Blue shift is similar to the red shift explained above for light leaving a neutron star, but this time caused by motion, not gravity.

The idea of red shift and blue shift caused by a star's motion is quite simple, and is based on the Doppler effect, which most people are familiar with. The classic example of the Doppler effect is the change in the pitch of a siren as an ambulance speeds past. The siren is more high-pitched than normal when the ambulance is approaching, and is more low-pitched once the ambulance has passed and is receding. The situation with HDE226868 is similar. Like sound, light travels as waves, but in this case the frequency of the waves determines the colour, not the pitch, of the light. In the part of HDE226868's supposed orbit in which the star is moving towards Earth, the light coming from the star will have a higher frequency than normal – it will be shifted towards the blue end of the spectrum. This is the equivalent of a higher pitch in the case of the ambulance. When the star is moving away from Earth, its light is red-shifted.

When astronomers examined the light coming from HDE226868, they found that it did indeed vary from being red-shifted to being blue-shifted – over a regular period of just 5.6 days. The brevity of this orbital period meant that the star must be orbiting something very massive. The astronomers could also discern a slight variation in brightness as the star orbited, too. They attributed this to the fact that the star was elongated, and not spherical. The reasoning behind this is that an elongated star side-on presents a greater surface area and will therefore appear brighter than the same star seen with its bulge facing towards or away from us. This fact led the astronomers to suppose that

HDE226868 was being stretched – by an intense gravitational field. This lent more support to the idea that the object around which the star was moving was a black hole.

Astronomers turned next to the X-ray emissions from Cygnus X-1. They observed how quickly these emissions varied, and found regular variations over a period of a few thousandths of a second. To an experienced astronomer – and for reasons that are not relevant here – this means that the object must be much smaller than any ordinary star. It had to be a very compact object indeed. Putting all the evidence together: there is a massive, compact object that gives out no visible light of its own, but does emit X-rays – presumably from its accretion disc. Cygnus X-1 is indeed a hot candidate for a black hole. Estimates put the mass of the invisible object in Cygnus X-1 at between eleven and twenty-one times the mass of the Sun. If this is true, and if general relativity is correct, it must be a black hole.

The main reason for the large uncertainty in the mass of the heavy object in Cygnus X-1 – between eleven and twenty-one solar masses – is that astronomers cannot tell at what angle the binary system is orbiting. From Earth, even the most powerful telescopes cannot separate the two objects. The light and X-ray emissions appear to originate in the same point in the sky. So all the information about the binary system was inferred from measurements of the variations in the X-ray and light received. The orbit might be nearly side-on; it cannot, however, be totally side-on, because in that case the two objects would eclipse one another. This would result in a different 'light curve', a graph of the light level over a period of days. Astronomers would be able to tell from the light – and the X-ray signals – that the two objects were passing behind each other. The opposite extreme, in which the orbit is inclined at 90 degrees to the Earth, is also ruled out, because in that case there would be no Doppler effect since the star would not be moving towards and away from us. Despite the uncertainty in mass, the fact that the lowest possible mass was well over three times the mass of the Sun meant that Cygnus X-1 was very exciting for astrophysicists.

In 1974 two celebrated black-hole theorists decided to have a bet on whether the source of the X-rays coming from Cygnus X-1 was the accretion disc around a black hole. They were Kip Thorne and Stephen Hawking. Both these scientists believed that black holes do exist. Stephen Hawking, of the University of Cambridge, explains why he went ahead with the bet: 'It was not because I didn't believe in black holes. Instead, it was because I wanted an insurance policy: I had done a lot of work on black holes, and it would all have been wasted if it had turned out that black holes didn't exist – but at least I would have had the consolation of winning the bet.' After enough evidence piled up in favour of Cygnus X-1 being a black hole, Hawking decided to concede the bet. While he was in Los Angeles in 1990, Hawking and his friends broke into Thorne's office, where the hand-written bet was hanging, on the wall, in a frame. Hawking put his thumbprint on the sheet of paper as a signature, and made sure that Thorne received his winnings: 'I had given Kip Thorne a year's subscription to *Penthouse*, much to his wife's disgust.' Hawking's thumbprint has come to symbolize the first real acceptance of the existence of black holes by the astrophysics community.

Black catalogue

Since the discovery of Cygnus X-1, several other very promising black hole candidates have been discovered and analysed, and astronomers have grown increasingly confident that they have definitive evidence of black holes. V404 Cygni is perhaps the most compelling case to date. As explained above, Philip Charles spotted the fact that V404 Cygni is an X-ray source, by looking at the data collected by the Ginga satellite in 1989.

Charles is a black-hole hunter based at the University of Oxford. His quest takes him to powerful telescopes all over the world: the Canary Islands and Hawaii in the northern hemisphere, and South Africa, Chile and Australia in the southern hemisphere. 'Searching for these things is the most wonderful way of going to the frontiers of modern physics … it's one of the most exciting things a modern scientist can do.' We have seen the lengths to which astronomers must go to analyse their quarry once they have found it – but finding it in the first place is not easy, in a deep sky filled with all manner of other objects. 'When you are looking for a needle in a haystack, you need the needle to shout out and say, "I'm here",' says Charles.

V404 Cygni did shout out, by way of the burst of X-rays detected by Ginga. Since 1989 Charles has been studying this binary system from the ground, at the William Herschel Telescope on La Palma in the Canary Islands. By studying the light and the X-rays from V404 Cygni, Charles and his colleagues concluded that the invisible object in the system must have a mass of at least six times that of the Sun, and probably more like twelve. This is why they are confident that this object must be a black hole. Presently, there are about twenty-five candidates for stars that have become black holes, including Cygnus X-1 and V404 Cygni.

Another member of the black hole hall of fame is GROJ1655-40. This binary system is 10,000 light years away, and appears in part of the sky in the constellation of Scorpio. In 1999 a team led by Rafael Rebelo at the Institute of Astrophysics on the Canary Islands made a remarkable finding related to this black hole candidate. They attempted to determine the gases around the two mutually orbiting objects that make up GROJ1655-40. If there really was a black hole here, it would have been formed from a massive star that would have been through the supernova stage. Would there be any record of a supernova in the gases in that region of space? What they found is remarkable: there were large amounts of the elements oxygen, magnesium, silicon and sulphur – which can only have been produced by a supernova.

However, while the evidence in favour of black holes is becoming ever more conclusive, certain problems still remain. One of them is that the X-ray radiation received from the black hole candidates is lower than that predicted by the theory of the accretion disc from which it emanates. In 1997 Dr Ramesh Narayan of the Harvard-Smithsonian Center for Astrophysics in Massachusetts proposed a theory that could solve the problem: he suggested that the reason that the amount of X-ray radiation is less than expected is that the black hole sucks the gas away inside its event horizon, before it can emit the radiation, so taking its energy with it. This requires an alternative to the generally accepted theory of the formation of the accretion disc. Narayan proposes that rather than forming flat, spinning discs, the gas from the stellar partner of the black hole or the neutron star falls in from all around. This would make it less dense, which would mean that it would retain its energy for longer – until it either hits the surface of the neutron star or moves within the event horizon, never to be seen again. In the case of a neutron star, the gas would still be able to radiate – since there is no event horizon, radiation can still be emitted into space.

To test his idea, Narayan looked at the radiation coming from eleven X-ray binary systems. Six of the systems – including V404 Cygni – were black-hole suspects, while the remaining five were thought to be neutron stars. And what he found is compelling: all six black-hole candidates were lacking in X-rays and the neutron stars were not. Narayan believes that if his theory is correct, then it will provide direct evidence of the existence of the event horizon. It may sound like Narayan has simply made up his theory to fit the facts, but the X-ray spectrum that his theory predicts matches what is observed extremely well.

The central idea
The black holes so far discussed are 'stellar' black holes which, as their name suggests, are created from dead stars. And they are all inside our own Milky Way galaxy. But the matter to make a black hole does not have to have come from a star – anything will do, as long as there is enough of it. For many years, astronomers have wondered whether there might be 'super-massive' black holes lurking at the centres of galaxies. There is enough matter in the centre of a galaxy to form a black hole of immense proportions.

The idea originated as an explanation of the behaviour of so-called 'active galaxies' – a class of deep-space objects, far outside the Milky Way. The first of these active galaxies was discovered, by American astronomer Carl Seyfert, in 1943. Seyfert galaxies, as they are now called, are large spiral galaxies that have surprisingly bright central regions. Discoveries of other types of active galaxies soon followed: radio galaxies in 1946, quasars in 1963 and blazars in 1968. Radio galaxies are huge: they have vast lobes in which electrically charged particles travel at close to the speed of light, emitting radio waves as they do so. The lobes can stretch thousands of light years into space. Quasars (quasi-stellar objects) are incredibly far away – some of them are the most distant objects ever detected. They are also the most powerful objects known, releasing phenomenal amounts of energy. Blazars are similar to quasars: in fact, they may be quasars that point directly towards us.

The first suggestion that super-massive black holes might be the source of power in active galaxies came in 1964, from American astrophysicist Edwin Salpeter at Cornell University and Yakov Zel'dovich in Moscow. Detailed theories to explain the processes behind the production and release of energy from super-massive black holes were developed through the 1970s, and the first strong evidence of their existence came in 1978, as a result of observations made by Wallace Sargent and Peter Young, based at the California Institute of Technology. These two researchers had turned their attention to a galaxy called M87, about 50 million light years away.

Sargent worked out how fast the stars in M87 are revolving around the centre, by studying the red shift and blue shift of starlight coming from around the galaxy's centre; Young estimated the mass of stars in the central part of the galaxy, and therefore the density of matter there. Both of the results suggested that there is as much mass concentrated in the centre of M87 as in about 3,000 million stars with the mass of the Sun – a black hole almost certainly exists there. In 1994, the Hubble Space Telescope was used to look at M87. The images produced were clear enough to show a central disk 500 light years across, made largely of hydrogen gas. Like a CD, there was a 'hole' in the middle of the disc, about a hundred light years across.

The Hubble Space Telescope provided evidence of another super-massive black hole in 1992, when it captured stunning images of a radio galaxy called NGC4261, also a member of the Virgo Cluster. Although NGC4261 is more than 40 million light years away, the Hubble Space Telescope produced a remarkably clear image of the centre of the galaxy, which showed a brown spiral disc inside the galaxy's white fuzzy core. The disc is rotating, and from the speed of the rotation astronomers determined the mass of the small central part of the core. They found that it has a mass more than 1,000 million times that of the Sun – fairly conclusive evidence of a super-massive black hole.

The Hubble Space Telescope has found other examples of what must be super-massive black holes, but perhaps the most compelling evidence came from the Very Long Baseline Array – a collection of radio telescopes across the USA. In 1994, astronomers used the array to examine a galaxy called NGC 4258. They collected radio waves emitted by water molecules near to the centre of the galaxy (as explained in 'The Rubber Universe'), and concluded that the density of the matter at the centre is more than 10,000 times greater than any known star cluster – there is almost certainly a black hole there.

Astrophysicists have long suspected that there may be a super-massive black hole at the centre of our own galaxy. The Milky Way is a spiral about 100,000 light years in diameter, containing about 100,000 million stars. Our Solar System lies about 25,000 light years from the centre of the Milky Way, in one of the spiral arms. If you find the constellation of Sagittarius, you will be looking in the direction of our galactic centre. For decades, it has been known that there is a massive object at the centre of our galaxy. The galactic centre is largely obscured from the view of optical telescopes by interstellar gas and dust, but radio telescopes detected a powerful source of radio waves in the 1950s, named Sagittarius A*. In the past few years, there has been a flurry of activity by astrophysicists trying to find out more about what it might be.

In 1997 astronomers at the European Southern Observatory used the New Technology Telescope to 'weigh' the huge object at the centre of our galaxy. They followed the stars for a year, tracking their motion – which is very slight from our distant viewpoint – and their results suggested that there is an object with a mass about 2.5 million times as much as the Sun, contained in a space less than one tenth of a light year across at the centre. This is strong evidence indeed to support the idea that there is a super-massive black hole on our galactic doorstep. There is no need to worry about this black hole: however hungry this object might be, we are sufficiently far away, and revolving around the galaxy at high enough speeds, to evade capture.

Perhaps in the future, intrepid astronauts might venture to take a closer look at the object at the centre of our galaxy. Given its extreme size and mass, would this be more dangerous still than a visit to a stellar black hole? Perhaps: near to the centre of a galaxy, stars are ripped apart to feed the huge black hole. Martin Rees takes us on a journey to the centre of the Milky Way: 'If one was to take a journey starting in the outer part of the galaxy and moving in, you'd see very hot glowing gas filling the galaxy, which is being heated by some mysterious powerful central source. You'd find the gas orbiting the disc, swirling at near the speed of light and getting very hot – so hot that it radiates not only light, but very energetic radiation like X-rays.'

Staying at a 'safe' distance and sending a space probe towards the centre, we would watch as the probe descends towards the event horizon, as in our imaginary journey to a stellar black hole. In fact, it turns out that this time the tidal forces are actually less severe: because the black hole is much larger, the curvature of space-time near to the event horizon is much less. The probe would survive passing through the event horizon, and, from its perspective, the light from outside the event horizon would be warped into a smaller and smaller space above the probe. There would be a final crunch as the probe met its fate at the singularity, but no one has any idea what might happen then – and Einstein's general relativity can provide us with no clue.

In addition to stellar and super-massive black holes, astrophysicists have recently discovered what they think are 'middleweight' black holes. In 1999 teams at Carnegie Mellon University and at NASA independently found evidence of this new class of black hole, in the spectrum of the X-rays emitted by the accretion of matter. The objects have masses of between 100 and 10,000 times that of the Sun – an intermediate size between stellar and super-massive. The researchers believe that these newly discovered objects may be the result of a kind of coalescence of many stellar black holes or neutron stars. The black holes they observed were in a 'starburst' galaxy – one in which star formation occurs at a phenomenal rate. Star death also happens quickly in this type of galaxy. The researchers estimate that millions of neutron stars and black holes would have been created in the galaxy they studied during the past 10 million years or so. Perhaps black holes – far from being elusive and rare – are actually extremely common. Each galaxy may have a super-massive one at its centre, many smaller ones forming all the time, and also a collection of intermediate ones.

Going for a spin

There is an important feature of black holes that has not been mentioned so far: the fact that nearly all of them are likely to be rotating. Certainly the super-massive ones at the centres of spiral galaxies will be, since spiral galaxies themselves are rotating. Stars rotate, too, and so when they become stellar black holes they will retain their spin. In fact, in the same way as a dying star gains speed as it shrinks to form a neutron star, a black hole is likely to be spinning much faster than the star from which it formed.

The rotation of a black hole has many interesting possible consequences. One of the most exciting, bizarre and controversial is the possibility of near-instantaneous travel into a rotating black hole and through a 'wormhole' to another point in space-time. The idea goes something like this: the rotation of a black hole may turn the singularity from an infinitesimal point into a ring; the 'well' of space-time inside a black hole is infinitely deep – open-ended – and so could conceivably connect to a similar well elsewhere. Because the singularity is enlarged, you could pass through it without being squashed to nothing. So you could travel through the wormhole and emerge from a 'white hole' at another point in space and time.

However, this theory has now been discounted. Kip Thorne: 'I recall a movie in which the spaceship goes inside a black hole and comes out the other side into another universe. That don't happen ... The fundamental laws of physics say that at the centre of a black hole, there is a singularity – a region in which space and time are infinitely warped, a region where when matter goes into it, matter is destroyed.' And Professor Roger Penrose of the University of Oxford says that 'these things are driven by romantic science fiction ideas ... The scientific evidence is that this is just not going to happen – it's just not possible.'

However, the idea of travel through wormholes may not be dead. Several theoretical physicists, Kip Thorne among them, have developed theories involving wormholes that would have no event horizon around their entrances and no singularities. So, these wormholes would connect objects that are not true black holes, but have similar properties. According to theory, these wormholes would collapse in on themselves unless they were 'held open' by some kind of exotic matter. This matter is as yet undiscovered, but its existence is predicted by various theories of modern physics. Some physicists believe that, in theory at least, such wormholes would allow instantaneous travel across space-time. Russian physicist Igor Novikov believes that travel across huge distances or through time would indeed be possible. Novikov says you would need to create two interconnected objects like black holes, and move them to two different locations: 'One of them can be near our Earth, the other in another galaxy, but this tunnel can be incredibly short – say a few metres … You can travel from one galaxy to another in, say, a few seconds.'

In 1997 a team at the Massachusetts Institute of Technology and the University of Maryland, led by Dr Shuang Nan Zhang, found a way to measure the spin of a black hole. What they did was to ascertain the distances from the centres of several black holes at which matter could attain a stable orbit. According to general relativity theory, matter that orbits beneath this distance is quickly sucked into the black hole, however fast it is moving. This limiting distance is farther from the centre of a black hole than the event horizon, and varies with the black hole's speed of rotation: the distance from a black hole of its 'last stable orbit' increases with the black hole's rotation.

The result of this is that the accretion disc of a black hole ends farther from the centre of a rotating black hole than from the same black hole if it were stationary. This should have the result that less X-ray radiation is emitted, since material would be able to radiate more energy before being sucked in if the accretion disc is closer to the black hole. Shuang Nan Zhang and his team managed to measure the last stable orbits of several stars, by examining the spectrum of X-rays emitted by the accretion disc. They found that two of them seemed to be rotating much more rapidly than the others. Interestingly, these two fast-spinning stars were also the only two that produced high-speed jets of material. Whatever the mechanism behind the jets, it probably has something to do with the black hole's spin. And so the fact that they occurred only with the two black holes that Shuang Nan Zhang believed were fast-spinning does suggest that his theory might be correct – although more work will need to be done to confirm this.

Theory tells us that there are only three things that we can determine about a black hole: its mass, its rate of spin and its electric charge. The mass can be worked out using the Doppler effect to measure the speed of a black hole's visible partner in a binary system, and Shuang Nan Zhang's method above seems to reveal a black hole's spin. So if astrophysicists can work out a way to measure the electric charge of a black hole, they would be able to know absolutely all of the information there is to know about a particular black hole. This would be satisfying in one way, but utterly frustrating in another. We would still be utterly confounded by the mysteries of what kind of star the black hole originally was, what kind of matter the hole has swallowed and what is going on inside the black hole's event horizon.

A rotating black hole literally drags space-time with it – rather like the way a whirlpool drags water around with it. This creates a region around the 'equator' of the black hole called the ergosphere. Travel within the ergosphere, and you will be dragged around with the black hole. Roger Penrose has devised an interesting thought experiment, in which it is possible to extract energy from a rotating black hole. If a piece of matter breaks up into two within the ergosphere, it is possible for one piece to gain speed by 'borrowing' energy from the other, and to shoot out of the ergosphere. The other piece will be swallowed by the black hole. Penrose himself explains the notion behind the 'Penrose process': 'You could imagine firing a number of particles or bodies into the neighbourhood of the spinning black hole, in the right direction ... one particle carries negative energy into the hole, and that means that the partner has more energy than the one that fell into it. So you can actually extract energy from a black hole using this process.'

Penrose can even see how the idea could be used as a huge power station at some point in the distant future. Penrose's power station is situated at a safe distance from the black hole, outside the ergosphere. The idea is to throw items of rubbish from there into the ergosphere, and as they break up, some are flung out at high speed, taking with them energy from within the ergosphere. The energy of these fast-moving objects could be used to drive some sort of turbine, to generate huge amounts of power. The loss of energy means that the black hole slows down, but so very slightly that you could go on using a black hole as a power station in this way perhaps until the end of time.

Penrose's scheme will never actually be realized: it is just a thought experiment to illustrate one of the strange aspects of black holes. A similar process – in which pairs of objects divide, one going into the black hole and one leaving – is Hawking radiation. This involves a concept of modern physics which, to the uninitiated, is hard to contemplate, but which to a modern physicist is as inevitable as black holes: virtual particles. While black holes are the domain of general relativity, which is used to study very large objects and their effects on space-time, virtual particles spring forth from quantum theory – the branch of modern physics that deals with very small objects such as atoms and subatomic particles.

One of the mainstays of quantum theory is the uncertainty principle – the idea that it is impossible to measure both the position and the speed of a particle with total certainty. One consequence of the uncertainty principle is the fact that there is no such thing as nothing: that even in a total vacuum, pairs of particles fleetingly come in and out of existence. These virtual particles normally annihilate each other within a tiny fraction of a second, and it is therefore not possible to detect them. However, the rate of generation of virtual particle pairs increases as space-time becomes more warped – such as near to a black hole. In 1974 Stephen Hawking showed that at the event horizon of a black hole, pairs of particles would be produced – from nothing – at a phenomenal rate. Some of the pairs would be annihilated, but others would be separated – one particle being sucked into the clutches of the black hole and the other moving off into space.

If virtual particle pairs are separated, each member becomes a real particle. One member of each pair of virtual particles has 'positive energy' and the other has 'negative energy'. The particle with negative energy goes through the event horizon and takes energy from the black hole to 'justify' its existence. The particle with positive energy will shoot off into space, and – according to Hawking's calculations – should be detectable. And so Hawking predicted that black holes should be radiating particles that have been created from nothing, as a result of the uncertainty principle. This whole process may sound unlikely, but has become part of the accepted theory of black holes.

The result of Hawking radiation on a black hole is that the hole gradually loses energy, which radiates away into space. Hawking viewed this as a kind of 'evaporation' of black holes. The rate of black hole evaporation should increase, and the size of a black hole decrease, until the black hole goes out of existence in a flash of energetic gamma rays. The theory is watertight, but no Hawking radiation has yet been detected. This is largely because black holes take in more mass and energy than they radiate, and so are growing, not shrinking. The intensity of Hawking radiation emitted by a large black hole is tiny compared with the radiation from its accretion disc, so the theory is hard to verify in practice.

However, another prediction made by Stephen Hawking concerns tiny black holes left over from the early stages of the Universe, when temperatures and pressures were high enough to cram huge amounts of matter into tiny spaces, smaller than an atom. Given the age of the Universe (see 'The Rubber Universe'), most of these tiny, 'primordial black holes' will each have exploded in a burst of gamma radiation by now. But Hawking has calculated that there may be 300 of them in each cubic light year. When they do explode, it should be possible to detect the gamma radiation produced, and some high-energy physics projects, including the space-based Compton Gamma Ray Observatory, are actually looking for them.

Hawking's theories of radiation and mini-black holes are the result of his attempt to reconcile general relativity and quantum theory – the physics of the very large and the very small. The combination of general relativity and quantum theory is called quantum gravity, and is one of the main preoccupations of modern theoretical physics. Both general relativity and quantum theory are well established, and their results are consistent and have been proved time and time again. For Kip Thorne, a successful theory of quantum gravity would allow physicists for the first time to understand the most elusive and worrying aspect of black holes: 'The singularity is an object at the core of a black hole that is governed not by the ordinary laws of physics, but by the laws of quantum gravity, which we're only just beginning to understand.'

The hunt for more black holes of all sizes continues. The recent discoveries of objects that seem to be black holes have spurred on Philip Charles, one of the most avid black-hole hunters: 'We need to get maybe twenty or thirty so that you can really see the population of black holes we are looking at: do they all have similar masses? Is there a spread of mass? Do they all come from the same sort of star? We can start to work that out once we can get enough of them ... There are things out there that are incredibly compact and they have to be explained; I think that's an exciting prospect.'

THE RUBBER UNIVERSE
... searching for the beginning of time...

The Universe is everything there is: space, time and 100,000 million galaxies. It is larger than we can truly appreciate, and it is growing. The space between galaxies is actually stretching, and the galaxies are flying apart at a phenomenal rate. This suggests that, long ago, the entire Universe was contained in a tiny space that exploded, and has been expanding ever since – this is the theory of the Big Bang. By working out the rate at which the galaxies are moving away from each other, astronomers can make a guess as to when the Big Bang occurred and therefore just how old the Universe is. This sounds straightforward, but is causing problems within the space science community: there is rivalry between astronomers who have calculated different speeds of expansion – but, what is more, some of the most reliable measurements suggest that the Universe is younger than some of the stars in it.

Pulling punches
The person who discovered that the Universe is expanding was American astronomer Edwin Hubble. He studied mathematics and astronomy at the University of Chicago, and also showed great promise as a boxer. But when he graduated in 1910 he put all pursuits aside to concentrate on becoming a lawyer. The fact that he decided to move back

into astronomy was to have a huge effect on the world – Hubble changed our view of the Universe for ever. In 1913, at the age of twenty-four, he began studying for a PhD in astronomy, at the University of Chicago. After finishing his studies in 1917 he served in the army during World War I, and then began working at the Mount Wilson Observatory, just outside Pasadena, California. It was there that he made his startling discoveries.

At the time Hubble was completing his PhD, the view of the Universe was already undergoing dramatic re-evaluation. The story begins with another American astronomer, Henrietta Leavitt. In 1907, working at Harvard College Observatory, Leavitt began studying a class of stars called Cepheid variables. These stars vary in brightness over a regular period of days or weeks. Through her painstaking work, Leavitt discovered that the period over which a Cepheid variable star varies depends on its average brightness. Astronomers measure a star's brightness in terms of its magnitude: the brighter a star appears, the lower its magnitude. A star of magnitude 6 is just visible to the unaided eye on a clear night, while the brightest star in the night sky, Sirius, has a magnitude of minus 1.5. During the latter part of the nineteenth century, the measurement of stars' magnitudes had changed from guesswork to a precise science, thanks largely to the introduction of photography.

Astronomers knew that if you could work out a star's actual output of light – its intrinsic luminosity – you could work out how far away it is from Earth. The reason for this is simple: two stars with the same intrinsic luminosity at different distances from Earth would have different magnitudes. In particular, if one star were twice as far away as another, it would appear only one-quarter as bright; three times as far, and it would appear one-ninth as bright, and so on. This is called the inverse square law, since how bright a light source appears depends upon the square of its distance from you. So, move a star four times as far away, and it will appear one-sixteenth as bright.

Alternatively, if two stars were at the same distance but differed in their intrinsic luminosities, they would again have different magnitudes. In this case, a star with twice the luminosity of another would appear twice as bright. However, stars are not all at the same distance, and they do not all give out the same amount of light. What was needed by astronomers trying to work out the distances to the stars was a 'standard candle' – a type of star whose intrinsic luminosity could somehow be ascertained with confidence. Then, by comparing this intrinsic luminosity with the actual magnitude as seen from Earth, a star's distance could be worked out using the inverse square law. Cepheid variables turned out to be just the standard candle that astronomers had been looking for.

Cepheid variables are named after the star 'delta Cephei', which is in the constellation Cepheus. This star is the 'prototype' Cepheid variable – it was the first such star to be studied in detail. Its magnitude varies from 3.6 to 4.3 and back to 3.6 again over a precise period of 5.4 days. Cepheid variables that are brighter than delta Cephei vary in magnitude over a longer period; less bright, and the period is shorter. Leavitt studied around thirty Cepheid variables in the Magellanic Clouds – large fuzzy areas in the sky, visible only from the southern hemisphere. Astronomers realized that all the stars in each of the two Magellanic Clouds were at about the same distance from Earth, but nobody knew how far. Henrietta Leavitt's work showed that there was a definite relationship between the intrinsic luminosities of these stars and the period over which their output varied, but until astronomers could work out how far just one of these stars was, this information was useless.

It was another American astronomer, Harlow Shapley, who worked out the distance to the Magellanic Clouds, and therefore 'calibrated' the Cepheid variables. Shapley was keen to find a way of measuring distances to the stars, particularly because he was troubled by the picture of the Universe as it stood in the early part of the twentieth century. Since the sixteenth century, when Copernicus realized that the Earth orbits the Sun, astronomers had assumed that the Solar System was at the centre of the Universe. Moreover, since the middle of the nineteenth century, astronomers believed that the Universe was about 23,000 light years across and about 6,000 light years thick. This had been worked out by looking at the distribution of stars in the sky; but Shapley was unconvinced. He was particularly troubled by the fact that the misty band of light called the Milky Way was seen only across part of the sky: surely if the Sun and the Earth were at the centre of the Universe, the sky should look the same in all directions.

In addition to the uneven distribution of the Milky Way, there were other problems for the accepted picture of the Universe. One of them involved globular clusters, which were known to be groups of thousands or millions of stars. Like the Milky Way itself, globular clusters were distributed unevenly across the sky, as seen from Earth. By studying Cepheid variables in the globular clusters, Shapley made a startling discovery: most of the globular clusters are tens of thousands of light years away. Together with the fact that they were found in only half of the sky, this challenged the idea that the Universe was only 23,000 light years across with the Sun at its centre. According to Shapley's observations, the Universe was almost ten times larger than people had believed, and the Sun was nowhere near the centre. This was big news, and Shapley became the most famous astronomer of his time.

The other problem for the picture of the Universe in the early part of the twentieth century involved the so-called 'spiral nebulas'. The word 'nebula' comes from the Latin word for 'misty', and was applied to any of many unidentified objects that appear in the night sky, most of them visible only through telescopes. In some, astronomers could make out individual stars, while others appeared as fuzzy, ill-defined patches of light. Some were irregularly shaped, some round, and some were the spirals that confounded the astronomers of the early twentieth century.

To some astronomers, the spiral nebulas seemed to be 'island universes', each a massive star system outside our own. Shapley, however, adhered to the idea that the spiral nebulas are inside our own system of stars, and that therefore the Universe – and not just the galaxy – is a disc about 200,000 light years across. Another American astronomer, Heber Curtis, believed otherwise: he was one of those who believed that the spiral nebulas are outside our own galaxy – although he stuck with the generally accepted idea that the Sun is at the centre of the galaxy. In 1920 Curtis and Shapley staged a well-publicized debate entitled 'The Scale of the Universe', at the National Academy of Sciences in Washington DC. Both men were right on one count and wrong on the other: we now know that the spiral nebulas really are other galaxies far outside our own, and that the Sun is indeed near to the edge of our own galaxy.

New reflections on the Universe
This is the point at which Edwin Hubble's contributions begin to be felt. From 1919 Hubble used a telescope with a

150-centimetre mirror to study the spiral nebulas, in an effort to settle the question of what these objects really are. He thought he could see Cepheid variables inside these nebulas, but could not be sure – he needed to use a more powerful telescope. So, between 1922 and 1924, Hubble used the largest telescope in the world at the time – the 250-centimetre Hooker Telescope, also at Mount Wilson. It had taken nearly six years to grind the concave mirror at the heart of the Hooker Telescope into the right shape. When, in 1930, Albert Einstein and his wife were being shown around the Mount Wilson Observatory, they were informed that the Hooker Telescope was the tool that astronomers were using to work out the nature of the Universe. Einstein's wife Mileva is reported to have remarked that her husband had done that on the back of an envelope.

In 1924 Hubble observed what was without question a Cepheid variable in the largest of the spiral nebulas – the Great Spiral of Andromeda, in the constellation of Andromeda. He charted the variation in the star's magnitude, and worked out that it had a period of one month. From this, he could work out the intrinsic luminosity and therefore the distance of the star. And from this, he could calculate the distance of the Great Spiral of Andromeda. His discovery was startling: he worked out that it is about one million light years away, much farther than Shapley's estimate for the size of the entire Universe. (It is now known that the Andromeda Galaxy, as it is now called, is more than two million light years away.) The conclusion was inescapable: the spiral nebulas were not nebulas at all, but were indeed star systems – galaxies outside our own. In 1926 Hubble devised a system of classification for the galaxies that he observed. He defined three types: spiral, elliptical and irregular.

At about the same time, the American astronomer Vesto Slipher was finding another vital piece of information about the spiral galaxies: the fact that nearly all of them are moving away from us. Slipher worked this out by studying the spectrum of light coming from each of them. The dark lines in the white light spectrum – described in Chapter Four in connection with the spectrum of sunlight – were all there, but were shifted somewhat towards the red end of the spectrum. The red shift of a galaxy is related to the speed at which it is moving away from us, in the same way as the pitch of a moving siren is lowered as an ambulance moves away. This is the Doppler effect, as explained in 'Black Holes'. If a galaxy is moving towards the Earth, its light is blue-shifted, so that the dark lines appear farther towards the blue end of the spectrum.

One of the galaxies that Slipher studied was the Great Spiral of Andromeda. Its spectrum is blue-shifted, and so it is coming towards us. The other galaxies seemed to be receding at phenomenal speeds. Slipher did not know how this fitted into the emerging picture of the Universe – but Hubble did. In 1927 Hubble looked at measurements of distance and red shift for twenty-four of the galaxies studied by Slipher. What he found was remarkable: the farther away a galaxy is, the faster it is receding. In 1929 he made a further discovery: if you divide a galaxy's speed by its distance from Earth, you will always end up with the same number. A galaxy twice as far is moving away twice as fast. The Universe is expanding in every direction.

The number obtained by dividing the speed of a galaxy by its distance from Earth is called Hubble's constant. Hubble's Law – as the relationship between a galaxy's distance and its speed is called – enabled astronomers to measure the distance to faraway galaxies in which individual Cepheid variable stars could not be seen. This was important, since only in those galaxies that lie relatively close can Cepheid variable stars be picked out. So, applying Hubble's law to more distant galaxies, with larger red shifts, astronomers began to realize that the Universe is much, much bigger than they had previously believed.

It would be easy to conclude from Hubble's Law that – since nearly all the galaxies are receding from us – the Earth has a special place in the Universe after all: at the centre. However, this is not the case. To see why, imagine a cake mixture with raisins evenly distributed inside it. In this analogy, the raisins correspond to galaxies, and the cake mixture corresponds to the space between them. Inside an oven, the cake mixture expands, pushing the raisins farther apart; the distance from one raisin to a neighbouring one increases. A raisin twice as far from another will be separated by twice as much of the mixture, and will therefore move twice as far in the same time. Applied to the Universe, the space between the galaxies is expanding. The Earth, the Sun and even the Milky Way are at no place in particular in a vast, expanding Universe.

New world view

Until the time of Copernicus, astronomers knew of five planets, which appear different from the stars because they shift their positions on a nightly basis relative to the star-studded backdrop. In fact the word 'planet' comes from the Greek word for 'wanderer'. When people aimed telescopes at the night sky, from early in the seventeenth century, it was inevitable that their ideas about the Universe would change radically for ever.

The magnification of a telescope is important when studying the Moon, the planets, comets and the nebulas (some of which were later identified as galaxies). But the light-gathering capability of a telescope is just as important, revealing objects much fainter than people can see with the unaided eye. In 1781 the English astronomer William Herschel discovered another planet – Uranus – which is not visible without a telescope. Herschel made many important contributions to astronomy, including being the first to attempt to work out the shape of the Milky Way. As a result of studying the distributions of stars in the night sky, he proposed in 1785 that the Milky Way is shaped like a lens – and he was not far off.

Long before Hubble's revelation that some of the nebulas are galaxies like our own, one of Herschel's contemporaries – German philosopher Immanuel Kant – proposed the same thing in 1755. But neither Herschel, Kant nor any of their contemporaries could unravel the nature of our galaxy or any other until astronomers had at their disposal powerful telescopes, a more sophisticated understanding of light and the technology of photography.

We now know that the Milky Way is a huge disc 100,000 light years across and 2,000 light years thick – not quite as large as Shapley had supposed – and that it contains about 200,000 million stars. Viewed from the side, it would appear as a long white streak, tapering at both ends and with a bulge in the centre. The bulge is about 6,000 light years across at its thickest point. About 200 globular clusters would be visible, dotted above and below the central bulge. Looking down on the disc of the galaxy, the central bulge would appear as a bright oval with long arms reaching out from it and trailing off into space to form a spiral shape.

The whole system is rotating, with the stars closer to the centre rotating faster than those further out – this is the origin of the spiral shape. The Solar System is in one of the

Milky Way's spiral arms – the Orion arm – about 23,000 light years from the centre of the galaxy. Although it is orbiting at a speed of around 240 kilometres per second, it takes about 200 million years to complete each orbit around the galactic centre. The Magellanic Clouds – where the Cepheid variables studied by Henrietta Leavitt are situated – are satellite galaxies, orbiting the Milky Way Galaxy once every 1,500 million years.

The Milky Way is the largest of a group of roughly thirty galaxies called the Local Group. Other members of this exclusive club include the Magellanic Clouds and the Andromeda Galaxy. The Local Group takes up about 125 million cubic light years of space. Travelling from Earth to the Andromeda Galaxy at the speed of, say, a rocket as it leaves Earth's atmosphere – 11 kilometres per second – this would take a staggering 70,000 million years. But the Local Group is just part of a larger cluster of galaxies, called the Local Supercluster. The centre of the Local Supercluster is nearly thirty times as far as the Andromeda Galaxy.

It is largely thanks to the efforts of Shapley, Slipher and Hubble that this picture of the Universe emerged. Cosmology – the study of the nature and the origins of the Universe – owes them a debt of gratitude. But there are many great minds working on the puzzling aspects of cosmology that is the legacy of these pioneers. The very relation that simplified our picture of the Universe – Hubble's Law – brought with it many questions, as yet unanswered. One of them is: 'How did it all begin?'

Hubble's constant – the number you arrive at when you divide a galaxy's speed by its distance from us – is given in units of 'kilometres per second per megaparsec'. A megaparsec is a unit of distance: one parsec is about 3.26 light years, so a megaparsec is about 3.26 million light years.

Hubble arrived at a value of the Hubble constant equivalent to 500 kilometres per second per megaparsec, meaning that a galaxy two megaparsecs (6.5 million light years) away is receding at a speed of 1,000 kilometres per second. Ever since the discovery of Hubble's Law, the numerical value of Hubble's constant has been the subject of debate and controversy. Incidentally, the reason that the Andromeda Galaxy is moving towards the Milky Way rather than away from it is that, being relatively nearby, the mutual gravitational attraction between the two is pulling them together. Only the more distant galaxies are moving away, and for them, Hubble's Law does seem to hold.

In the beginning

Hubble wondered whether the fact that the Universe is expanding means that at some time in the past it was condensed into a much smaller space – the concept we now know as the Big Bang. This idea had already been put forward in 1927, by a Belgian priest and astronomer, Abbé Georges Lemaître. Combining the findings of Shapley, Slipher and Hubble with Einstein's General Theory of Relativity, Lemaître gave the idea of an expanding Universe a firm mathematical footing, showing how it could be explained by the expansion of space, as illustrated by the rising cake analogy above, and not in terms of the galaxies physically flying apart.

Using Hubble's constant – a key to the rate of expansion of the Universe – successive generations of astronomers have worked backwards to estimate how long the Universe has been expanding, and therefore how old it is. Using his own calculation of Hubble's constant during the 1930s, Hubble worked out the age of the Universe to be around 2,000 million years. Geologists had already worked out that the ages of some rocks found on Earth were much older than this, and so Hubble's calculation caused much consternation in the scientific community as a whole.

The Big Bang theory was not the only one that was proposed to explain the creation of the Universe. Its main rival was the Steady State theory. This was put forward by a trio of mathematicians and astrophysicists, Fred Hoyle, Hermann Bondi and Thomas Gold, in the 1940s. It too appealed to general relativity. These three scientists considered it ludicrous that anything could be created by an explosion. However, the only way their theory could account for the fact that the Universe is expanding was to suppose that new matter was being created from nothing all the time. Still, both the Big Bang and the Steady State theories had their supporters, and fierce debates raged.

Also during the 1940s, Russian-born American physicist George Gamow improved the Big Bang theory, proposing that the initial state of the Universe would have been an intensely hot, extremely compact 'fireball' smaller than an atom. Gamow also showed how the Big Bang idea could explain the production of hydrogen and helium in the early Universe. Professor Alan Guth at the Massachusetts Institute of Technology points out that this is one of the strong points of the Big Bang theory: 'The Big Bang theory has to be recognized as being more than just a cartoon image of an explosion from which the Universe came: it really is a detailed mathematical theory. And given this mathematical description of the details, you can actually calculate the rates of different nuclear reactions that would have taken place in the early Universe.'

By the 1950s, the idea that the Universe was once searingly hot and incredibly dense began to be generally accepted by most cosmologists. But the really conclusive evidence in favour of the Big Bang came, by accident, in 1964. American radio astronomers Arno Penzias and Robert Wilson, working at the Bell Telephone Laboratories in New Jersey, were attempting to identify all sources of radio interference, so that communications with satellites could be optimized. When they had eliminated all the known sources of radio waves, they were left with one signal they could not, at first, explain. On further study, and after discussing their findings with Bernard Burke at the Massachusetts Institute of Technology, they realized that they had inadvertently discovered the radiation left over from the Big Bang.

The cosmic background radiation, as Penzias and Wilson's discovery became known, is a microwave signal. Just as the spectrum of light coming from a hot object can reveal that object's temperature – a white-hot object is at a higher temperature than a red-hot one, for example – radiation emitted by any object is characteristic of its temperature. The signal that Penzias and Wilson had detected corresponded perfectly to a temperature of minus 270 degrees C, just under three degrees Celsius above the coldest possible temperature, called absolute zero. Despite its intensely hot beginning, the Universe today is on average a very cold place indeed.

The temperature indicated by the cosmic background radiation fitted almost perfectly with one of the predictions of the Big Bang theory – namely that the intense heat of the early Universe would have produced high-energy radiation. As the Universe expanded, the wavelengths of the radiation would have been drawn out, gradually changing short-wavelength gamma radiation to the long-wavelength microwave radiation that Penzias and Wilson had picked up with their radio antenna. Another way of looking at this change in wavelength is that the Universe has cooled as it has expanded.

And so the Big Bang theory became the only real contender for explaining the creation of the Universe. The fact that the whole Universe began at one point and expanded in all directions means that it has no edge. Furthermore, if the Big Bang really was the moment of creation of space and time, then it would have occurred in every point of the Universe, not in any particular location. However, it is the space between the galaxies that is expanding, not the space within a galaxy or between galaxies that are near each other. Gravity is holding space together in these locations.

Bizarre though the predictions of the Big Bang theory may seem, the theory has been confirmed in many other ways since the 1960s. Perhaps the most convincing confirmation came from a space-based observatory called the COBE (Cosmic Background Explorer). Launched in 1989, COBE mapped the entire sky, making extremely precise measurements of the radiation left over from the Big Bang. Its triumphant year was 1992, when the map was complete. Alan Guth: 'It's in absolutely gorgeous agreement with the predictions of the thermal radiation from the heat of the Big Bang.' Perhaps even more importantly, there were very slight variations in the intensity of the background radiation, exactly as predicted by the Big Bang theory: if the radiation had been perfectly uniform across the entire sky, there would have been no way for the galaxies to have formed in the early Universe; the entire Universe would have been perfectly smooth.

Conflicting measurements
Edwin Hubble's original estimate of the Hubble constant put it at about 500 kilometres per second per megaparsec.

Although cosmologists agree that this is far higher than the actual value, there is still a great deal of disagreement about the actual figure. The values determined between the 1960s and 1990s generally fell into two camps. It was during the 1960s that the rivalry began: one team comprising Allan Sandage at the Carnegie Institution in the USA and Gustav Tammann at the University of Basel, Switzerland, found a relatively low value of 50, while Gérard De Vaucouleurs at the University of Texas and his supporters found a value around 100. Each set of investigators held on avidly to their results, and the debate still rages. The difference between 50 and 100 may not seem too great, but a value as low as 50 suggests that the Universe is perhaps twenty thousand million years old, while a value of 100 suggests nearer eight thousand million years.

Professor John Huchra, at Harvard University, makes it clear that the Hubble constant is very important: 'To a cosmologist, the Hubble constant is perhaps the most important number – in part because it sets the scale of the Universe; it sets the sizes of the things we look at; it sets the luminosities of the galaxies we look at; it sets the age of the Universe. We need to understand that if we really want to talk about the model as a whole and understand where we come from.'

Of the two measurements that must be made for accurate determination of the Hubble constant – the speed at which galaxies are moving away and their distances from Earth – the first is by far the easier. It is calculated using the Doppler effect. The distance of a galaxy is much more difficult to ascertain. The method using Cepheid variables is fairly reliable, but can be used only for nearby galaxies, because individual stars can rarely be observed in very distant galaxies. And, frustratingly, it is the most distant galaxies that are most important in determining the Hubble constant, since those will be the ones unaffected by the gravitational effect of our nearest neighbouring galaxies.

Several different methods for determining the distances to faraway galaxies – and therefore the Hubble constant – have emerged since Hubble first discovered the relationship between galaxies' speeds and their distances. Sandage and Tammann decided to study supernovas – bright outbursts produced during the death throes of aging stars. According to these two cosmologists, all supernovas produce approximately the same amount of light at maximum intensity. This is useful because, as Tammann points out, 'if you can calibrate this luminosity in some galaxy whose distance you know already through Cepheids, you have a standard candle'. Because supernovas are much brighter than Cepheid variable stars, they can be observed in galaxies that are much farther away. It was using this supernova method that Sandage and Tammann calculated their value for the Hubble constant as 50 kilometres per second per megaparsec.

However, there are problems with Sandage and Tammann's method. First, there is disagreement as to whether all supernovas produce the same amount of light. Second, supernovas are rare events. Sandage and Tammann base their measurements on a supernova that occurred in 1937. The intensity of this supernova was measured by astronomers Walter Baade and Fritz Zwicky, by analysing photographic plates that had recorded the outburst. Michael Pierce of Indiana University explains that he and his associate George Jacoby, at the Kitt Peak National Observatory, 'have been analysing the plates using modern techniques, and have found that the supernova was considerably fainter than Baade and Zwicky had claimed ... and this correction alone would revise the Sandage and Tammann value from about fifty to about sixty or sixty-five'.

Another, rather indirect, method of determining the Hubble constant involves measuring the rate of rotation of spiral galaxies. In a rotating galaxy, half the stars are moving towards us and the other half are moving away. Using the Doppler effect, it is relatively simple to work out how fast the galaxy is spinning and, since the rate of spin is related to the mass of a galaxy, the galaxy's mass can be estimated using this method. The mass of a galaxy is related to the number of stars, and the number of stars is related to the total output of light, and therefore the galaxy's intrinsic luminosity. By comparing the luminosity with the galaxy's magnitude, its distance can be worked out. This method tends to find values of the Hubble constant much higher than the value that Sandage and Tammann had worked out.

Michael Pierce, one of the astronomers using the rotation of spiral galaxies to calculate the Hubble constant, says of this technique: 'If one dismisses the idea that somehow the local galaxies are fundamentally different from the more distant ones – and I see no reason to assume such a thing – then I have to accept that the Hubble constant is eighty-five.'

Yet another approach is based on the fact that distant galaxies appear as relatively uniform patches of light, while those that are nearer appear 'grainy' because individual stars can be made out. In the same way, the individual dots that make up a newspaper photograph are visible only when seen close up. So, by studying the variation in light intensity across the surface of a galaxy, it is possible to gain an idea of how far away a galaxy is. This method, developed by John Tonry of the Massachusetts Institute of Technology, seems to be in good agreement with most of the other methods, but cannot be used with the most distant galaxies.

Apart from Sandage and Tammann's value of around fifty kilometres per second per megaparsec, all the other methods so far described have calculated the Hubble constant at higher values. These higher values suggest that the Big Bang happened around twelve thousand million years ago – and the very highest values suggest an even lower age for the Universe, just eight thousand million years. This is a problem for cosmologists, because some stars in the Milky Way Galaxy are known – with some certainty – to be older than that.

Theories of star formation and evolution are well developed, and probably very accurate – as Professor Pierre Demarque of Yale University in Connecticut points out: 'This is very mature physics, compared to what we use in extragalactic physics.' The oldest stars in our galaxy are found in the globular clusters, distributed around the galactic centre. Astronomers are able to work out the ages of globular clusters by examining the proportions of stars within them that have reached the late stages of development. But there are other ways of working out how old stars are.

Professor Michael Rowan-Robinson of Queen Mary and Westfields College, London, says, 'We can determine the ages of these stars not only by studying star clusters, as we do for the very oldest stars, but also by looking at the radioactive elements, similar to how scientists use carbon-14 dating ... Then there's a third method, which is based on white dwarf stars – which is what the Sun will end up as. Because we know how fast these stars cool, we can work out how old these stars are.' Professor Demarque adds another field of study to the list of reasons why astronomers are fairly confident about their understanding of stars: 'We have better ways to probe the interiors of stars; one of these is the new field of helioseismology, which enables us to study the interior of the Sun by studying its oscillation.'

So astronomers are fairly sure that their theories of star evolution are good enough to enable them to be confident that some stars in our galaxy may be as old as 14,000 million years. And yet this is at odds with the higher values for the Hubble constant, which put the age of the Universe as less than the age of the oldest stars. As George Jacoby of the Kitt Peak National Observatory says, 'That is a real conflict. How do we resolve this conflict? Well, I'm not sure how …'

Help is at hand – from Edwin Hubble's namesake, the Hubble Space Telescope. From the outset, one of its mission objectives was to attempt to settle the debate over the Hubble constant. A team of scientists working on the Key Project on the Extragalactic Distance Scale – in consultation with some of the major players in the debate – set out to determine the Hubble constant to within 10 per cent.

The approach taken by the Key Project team was to check galactic distances found by several different methods against the most reliable method – Cepheid variable stars, for as many galaxies as possible. The superior view of the heavens that Hubble gives astronomers meant that the Cepheid variable technique could be extended to more distant galaxies than was possible before. Cepheid variables in galaxies as far away as sixty-five million light years – in the Virgo and Coma galaxy clusters – were observed with great accuracy, to gauge their luminosities and therefore their distances from Earth.

By the time the Key Project came to an end in May 1999, the Hubble Space Telescope had peered at a total of almost 800 Cepheid variables in eighteen galaxies. The results were encouraging: the team had calculated a value for the Hubble constant of 71 kilometres per second per megaparsec, set pleasingly between the high and low values previously worked out, and leading to an age of the Universe that does not conflict with the ages of the oldest stars.

Far out

All the approaches to calculating the Hubble constant outlined so far – including the research carried out by the Hubble Space Telescope team – are only really useful for nearby galaxies or those at intermediate distances. What is needed for a truly accurate determination of this elusive and all-important number are the distances to the most remote galaxies – thousands of millions of light years away.

Two promising methods are beginning to yield interesting results for galaxies lying at these immense distances across space. The first of these is particularly promising, because it can gauge how far an entire cluster of galaxies is receding. This approach makes use of a phenomenon known as the Sunyaev-Zel'dovich effect. Every cluster of galaxies is shrouded in tenuous but extremely high-temperature plasma (ionized gas). The plasma can affect the cosmic background radiation passing through it. As the radiation meets the gas, it can receive extra energy from the electrons in the plasma, and this affects the spectrum of the radiation. Because the plasma is so tenuous, collisions of radiation and electrons are rare, so the effect is very slight – the radiation is predicted to change by only one part in 100,000.

The Sunyaev-Zel'dovich effect enables astronomers to estimate the size of a galaxy cluster: the greater the size of a cluster, the more chance that the radiation will collide with electrons in the plasma. This done, astronomers can measure how large the galaxy appears in their telescopes, and estimate how far away it is. A project run by NASA's Dr Marshall Joy and Dr John Carlstrom of the University of Chicago has been investigating the effect since 1992, using only microwave signals. More recently, they have been able to compare observations made using ground-based radio telescopes with observations from the orbiting X-ray observatory, Chandra, launched in 1999. Joy and Carlstrom's preliminary values of the Hubble constant are between 40 and 80 kilometres per second per megaparsec but the pair hope to narrow down the range of values.

The fact that light is bent by the gravitational fields around massive objects is the basis for another distant galaxy approach to the Hubble constant problem. As light passes through a large galaxy, it can be bent in the same way as light is bent by a lens. The distance between Earth and a lensing galaxy can be estimated from the effect the galaxy has on the light from even more distant galaxies. In particular, light from the more distant galaxies that takes different paths through the galaxy also takes different amounts of time to reach Earth. Astronomers can work out this difference if something changes in the more distant galaxy – such as a supernova.

Then, the same fluctuation in brightness will be observed at different times and, from this, astronomers can work out the distance of the lensing galaxy. The phenomenon of gravitational lensing is a prediction of general relativity, and is now well tested. But examples of gravitational lenses are few and far between, and using the phenomenon to determine the Hubble constant has yet to prove itself. Preliminary results again cover a range of values, but are pleasingly within the mid-range of values, around sixty-five kilometres per second per megaparsec.

The gravitational lensing and the Sunyaev-Zel'dovich effect seem to point to values for the Hubble constant that fall between the two extremes suggested by the earlier investigators. So did the research carried out by the Hubble Space Telescope's Key Project Team. However, just as everything seemed to be settling down in the debate, another method has found worrying evidence that the larger values of the Hubble constant might just be right after all.

In September 1999 the Hubble Space Telescope was turned to a galaxy named NGC 4258. This spiral galaxy had been used as a benchmark, because it was one of the farthest galaxies whose Cepheid variables could be seen clearly. It was thought that the galaxy was at a distance of about 8.1 megaparsecs (26.4 million light years) from Earth. With the clearest picture yet of the galaxy, Dr Eyal Maoz and his team at NASA's Ames Research Center in California was able to make what he thought was a reliable judgement of the distance of this galaxy, to confirm this figure. However, a new measurement of the distance of NGC 4258 was made using another, more reliable method.

Near to the heart of the galaxy, water molecules are heated by the release of huge amounts of energy by a supermassive black hole and give off microwave radiation at a very precise frequency. The radiation is created by a process much like the way a laser produces light. And just as the word 'laser' is derived from the acronym 'light amplification by stimulated emission of radiation', so the phenomenon observed in NGC 4258 is a 'maser', from the same acronym but with 'light' replaced by 'microwave'. The microwaves are produced in two jets that emerge from the galaxy's core.

By following the way the maser jets move, a collection of radio telescopes on Earth – called the Very Long Baseline

Array – was able to work out how large the galaxy is. Using simple geometry, based on how large the galaxy appears, astronomers could then work out how far away the galaxy is. Their calculations put the galaxy at a distance of about 6.1 megaparsecs (20 million light years). Comparing this highly accurate measurement of distance with Maoz's measurement based on Cepheid variables seems to highlight a flaw in the use of Cepheid variables. If this is correct, then all previous measurements using Cepheids may have been consistently too high by around 12 per cent. And if this is true, then all the previous measurements of the Hubble constant also suffer from that same flaw, since they have all ultimately been calibrated been using Cepheid variables. What this means to cosmologists is that the age of the Universe is about 12 per cent lower than the previous best estimates: perhaps as little as 12,000 million years.

This low value has resurrected the paradox that cosmologists feared when previously confronted with large values of the Hubble constant: stars in our own galaxy are known to be older than the age of the Universe that is suggested by these figures.

Fatal attraction

So at present the world of cosmology seems to be in disarray. Despite some of the most sophisticated tools ever constructed, and a highly developed understanding of the Universe, there is disagreement and paradox. And there is another problem: the effect of gravity. The mass of the Universe has the effect of slowing the rate of expansion, pulling everything together. The more mass there is, the greater this effect. This process is well understood – in terms of general relativity; the trouble is that cosmologists do not know how much mass there is in the Universe. If they did, they would be able to calculate how much the rate of expansion has changed since the Big Bang.

The situation is rather like watching an athlete at only one point during a race and trying to infer from that how long he or she has been running. You might be able to calculate how fast the athlete is running at that point, but he or she will not have been running at that speed for the whole race. It is likely that the speed was higher at the beginning of the race. If you knew at what rate the athlete had been slowing down, then you could produce a reliable estimate of when the race began.

In the same way, even if cosmologists could work out a reliable figure for the expansion of the Universe at the present moment – from an accurate determination of the Hubble constant – they would not necessarily know how old the Universe is. They need to know the extent of the slowing effect of gravity, and they can work that out only if they know how much mass is contained within the Universe. In other words, they can work out only how the expansion has slowed over time – and therefore how old the Universe is – if they can ascertain how much matter there is in the Universe.

According to Einstein's General Theory of Relativity, anything that possesses mass – stars and planets, interstellar gas and dust, and black holes, for example – warps space-time, the fabric of the Universe. This is explained in 'Black Holes'. Gravity between massive objects holds space-time together locally – in galaxy clusters, for example – but the space between clusters of galaxies is expanding steadily in all directions. This expansion is, of course, what the Hubble constant attempts to quantify. But as well as the local-scale distortion of space-time, matter is curving space-time on the scale of the entire Universe. Depending upon how much matter there is – and therefore to what extent the gravitational effects of that matter is pulling space-time back in on itself – the Universe is either 'closed', 'open' or 'flat'.

The behaviours of space-time and the objects within it are based on a branch of mathematics called non-Euclidean geometry. This is a description of space that is different from the descriptions and theorems of Euclid – one of the most influential and prolific mathematicians of ancient Greece. Around 300 BC, Euclid collected a number of mathematical theorems and put them together into a collection of thirteen books called the *Elements*. In an attempt to define fully the nature of shapes and the space they inhabit, Euclid included in the *Elements* a set of ten axioms and postulates, which were supposed to be self-evident truths. Using these as starting points, Euclid went on to set out 465 theorems that helped mathematicians to work out areas and volumes of shapes, for example. The *Elements* was the standard mathematical textbook in many countries of the world for more than 2,000 years.

A classic example of one consequence of Euclid's geometry is the idea that the angles in the corners, or vertices, of any triangle add up to exactly 180 degrees. This is easy to verify for a triangle drawn on a flat surface. But there is a problem if you draw a triangle on a curved surface, such as the surface of the Earth. Imagine a triangle with the North Pole as one vertex and two points on the equator as the others. Since the angle between any line joining the equator to the North Pole is at 90 degrees to the equator, and there are two such lines in this triangle, the angles inside the triangle will add up to more than 180 degrees. The angles inside the triangle that joins Kenya, Papua New Guinea and the North Pole, for example, add up to about

270 degrees. The important feature of this triangle, as opposed to those that Euclid was concerned with, is that it is drawn on to a curved surface rather than a flat one.

Another of Euclid's postulates is the idea that parallel lines will remain at a fixed distance from one another along their entire length. This sounds logical enough, until once again you consider a curved surface. Two people setting out from locations on the equator and travelling due south – in parallel with each other – will converge gradually, and meet at the South Pole. The relevance of this non-Euclidean geometry to the Universe at large is that gravity curves space-time, and therefore causes unfamiliar effects in three or four dimensions, just as these triangles and parallel lines do on a curved two-dimensional surface. As we have seen, for example, the fact that parallel rays of light passing either side of a massive galaxy converge – gravitational lensing – is understood in terms of the curvature of space-time.

The curvature of a surface may be 'open', 'closed' or 'flat'. A triangle drawn on a sphere – a closed surface – has angles that add up to more than 180 degrees. But on an open surface, triangles have angles adding up to less than 180 degrees. An example of such a surface is a saddle. In the same way, the curvature of space-time may make the Universe flat, closed or open. If there is enough matter in the Universe to curve space-time in on itself, then we have a closed universe. If not, then the Universe is either open or 'flat'. One of the main aims of modern cosmology is to discover just how curved the space-time of the Universe is. To do this, they must find out how much mass it contains and how big it is.

An open universe will go on expanding for ever, with the galaxies moving farther apart into eternity. This would be a slow, cold death for the Universe, and would result from there not being enough matter to halt the expansion. A closed universe is perhaps an even more frightening prospect: it will eventually stop expanding, and begin to be pulled in on itself, shrinking back at an ever-increasing rate, eventually reducing in size to nothing. Cosmologists have given this idea a name – the Big Crunch – to relate it to the Big Bang that occurred at the other end of the Universe's lifetime. The other possibility is that space contains just the right amount of matter to stop expanding, but not to contract again. In this case, the Universe would be described as 'flat', with what is called 'critical density'.

A handy analogy may help clarify these complexities: a part-inflated balloon with dots drawn on to its surface. In this analogy, the dots represent the galaxies, while the rubber itself represents the space-time they inhabit. Inflating the balloon makes it equivalent to an expanding Universe: the dots move farther apart in the same way as galaxies are observed to do in the Universe; space-time (the rubber) is stretching, as a result of the Big Bang. In this case, the three dimensions of space are confined to two dimensions in the surface of the rubber. The Universe has no centre, and no edges. Just as a traveller on the 'closed' surface of the Earth can journey around the world and return to where they started – a journey across a closed universe could conceivably bring you back to Earth, after a journey of trillions of years.

An 'open' universe does not contain enough matter to close space-time in this way, and neither does a 'flat' universe. To work out which of these geometrical descriptions applies to our Universe, cosmologists must work out the average density of space. To do this, they must be able to measure the volume of the Universe (or at least large swathes of it) and the total amount of matter contained within it. This sounds like a daunting task – and it is.

Dark secrets

At present, cosmologists have little idea of the total mass present in the Universe. What they do know is that the mass they can observe puts the density of the Universe at less than one-tenth that required to attain critical density (the amount of matter required to halt the expansion of space-time). And yet large-scale observations of the Universe tend to point to a flat universe, which means the density must be close to the critical density. And so there must be much more mass in the Universe than cosmologists can actually observe.

The measurement of the density of the Universe is important not only to predict the future, but to look into the past. The reason for this is that estimates of the age of the Universe (based on the Hubble constant) will depend upon the amount of mass in the Universe. If there is a great deal of mass in the Universe – if the density is high – then the expansion rate will be slowing down quickly. In that case, for the expansion rate still to be as high as it is today, we must be in a relatively young Universe. Conversely, if the density is relatively low, then the Universe must have had a longer time to reach this speed, and we are in a relatively old Universe. In this way, a low-density Universe would help to increase the age of the Universe as calculated using the high values for the Hubble constant found by most experimenters. This would be a relief, as it may just push the age of the Universe beyond the ages of its oldest stars.

It is relatively simple to estimate the overall mass of observable matter in the Universe, by measuring the total output of light from each galaxy, and taking into account the dark 'lanes' in galaxies due to interstellar dust. But cosmologists are sure that there exists a huge amount of matter that cannot be directly observed – what they have called 'dark matter'. Astronomer Heather Morrison at the

Kitt Peak National Observatory in Tucson, Arizona, explains the derivation of the term: 'When we are looking for that matter, we can't account for a significant amount of the mass, so we've had to postulate something we call dark matter to explain it. We call it dark matter because we've never observed any light from it. But it's worse than that – we don't actually know what it is.'

The first clues to the existence of dark matter came in the 1930s, when astronomer Fritz Zwicky studied the motions of clusters of galaxies. Zwicky worked out that, based on the amount of matter that could be seen in these clusters, the galaxies should be flying apart, not attracting one another as they are seen to do. The gravitational influence of some kind of unseen matter is holding these clusters together. Even the mutual gravitational attraction between the Milky Way Galaxy and the Andromeda Galaxy is far greater than can be accounted for by the amount of observable mass. Also, when observing distant stars within our own galaxy, gravitational lensing occurs without anything observable between the Earth and those distant stars.

There are several pet theories to explain the nature of dark matter. Two of them involve tiny particles that would have to exist in great numbers. Both of these types of particle would have to interact very weakly with ordinary matter – otherwise scientists would be able to detect them as they arrive in Earth- or space-based detectors. Two candidates are generally put forward: neutrinos – produced in certain nuclear reactions, and in particularly huge numbers during the Big Bang itself – and WIMPs (weakly interacting massive particles), which were also thought to be produced during the Big Bang. Because of their elusive nature, these particles have yet to be detected in the quantities needed to contribute sufficiently to the dark matter of the Universe.

The gravitational lensing effects of dark matter are blamed upon hypothetical objects called MACHOs (massive compact halo objects). These objects would be large and invisible, and would have high mass – the obvious candidates are black holes. And early in 2000, astronomers based in Australia and in Chile, having used ground-based telescopes as well as the Hubble Space Telescope, announced that they had made the most promising breakthrough in the search for MACHOs so far. The astronomers discovered what they are certain must be two isolated black holes in our own galaxy.

Until now, black-hole candidates had been discovered only by the X-ray emissions from the accretion disc that forms as gas from a binary companion star is pulled off by the black holes' strong gravitational field, as explained in 'Black Holes'. The individual examples were discovered by their gravitational lensing effects on more distant stars. In particular, the Hubble Space Telescope detected that light from distant stars became brighter and dimmer – presumably as it was focused by black holes passing in front of the stars.

The 'gravitational microlensing events' lasted several hundred days each, allowing telescopes based on the ground to study the effects over a long period, while the space telescope moved on to other duties. Detailed calculations highlighted the fact that the unseen objects must have a mass at least six times that of the Sun – too big to be a white dwarf or a neutron star. The existence of these two black hole candidates suggests that isolated black holes may be common in our galaxy – and other galaxies. Larger stars use up their hydrogen fuel more quickly than smaller stars, and it is the larger stars that, according to theory, go on potentially to become black holes. So, given the age of the Milky Way, it seems reasonable to expect that large numbers of massive stars have already burned themselves out and formed black holes. The astronomers who made the discoveries think that isolated black holes could account for a reasonable proportion of the hypothetical MACHOs in the Universe. This may be the first step towards tracking down the 'missing mass' of the Universe.

In one sense, cosmologists are keen to find the perceived missing mass of the Universe, so that the value they calculate for the density of the Universe moves closer to the critical density to which their observations lead them; on the other hand, they would rather have a much lower density, so that the age of the Universe calculated by the best values of the Hubble constant is pushed up somewhat.

The other crucial measurement required for a calculation of the density of the Universe is its size. All we have to go on for that measurement is the observable Universe – bounded by the most distant objects that we can detect. Until recently, the most distant objects observed were quasars – active galaxies whose red shifts are so extreme that they are estimated as being at least 12,000 million light years away. Since light takes time to cross such vast distances, this is not only a very long way away, but also a long way back in time – to the early stages of the Universe. But unfortunately, the accuracy of this estimate of distance is unknown: the measurement is based upon the red shift of these distant galaxies, using the best value for the Hubble constant. This value is in dispute, as we have already seen.

However, in 1999 the Compton Gamma Ray Observatory – an orbiting telescope that can see gamma rays from space – detected events that seem to be occurring at distances that are farther still, and scientists used a different method to calculate their distance. The events are known as gamma ray bursts, and just what causes them is not yet known. However, what is known is that the spread of radiation that we receive in our instruments in orbit around Earth is related to the intrinsic luminosity of the events – the actual amount of energy released. This is similar to the use of Cepheid variables in determining the distances to nearby galaxies.

The ultimate questions to which many people want answers are, 'What is outside the Universe?' and, 'What existed before the Big Bang?' These questions can be posed in another way: 'Where and when does the Universe exist?' Modern cosmology avoids these questions, arguing that time and space were created in the Big Bang – there simply was no time before the beginning of the Universe; and the Big Bang created all space, too, so there is no space outside the Universe. Professor Carlos Frenk of Durham University explains: 'Given that space and time are so closely related to one another, asking the question, "What was there before the Universe?" is exactly the same as asking, "What is there outside the Universe?" – no difference: space was created with the Universe, time was created with the Universe. Time is a concept that is part of the Universe: it did not exist before the Universe came into existence.'

In other words, the Universe is nowhere and 'no-when', since space and time are intrinsic qualities of the Universe itself. Most people are left unsatisfied with such answers, but cosmologists insist that their arguments are watertight, and that we must adjust our common-sense notions of space and time, as relativity theory has shown us.

For now, one of the main questions of cosmology is: 'What is the ultimate fate of our Universe?' We will be able to find an answer only by working out a value for the

Hubble constant as well as determining the density of the Universe as a whole. If there is enough matter in the Universe to halt the expansion and pull space-time in on itself, then it seems that it will all end as a super-dense fireball, just as it began. The Universe could end up as a tiny black hole – a singularity. If there is to be a Big Crunch, it will not happen for many billions of years. By that time, perhaps cosmologists will have agreed on a reliable value for the Hubble constant.

Another cosmological theory, called the Oscillating Universe Theory, says that the Universe could be created again from the singularity that results from the Big Crunch: there could be another Big Bang. So perhaps everything will simply start again.

The Universe is everything there is: space, time and a hundred thousand million galaxies. It is larger than we can truly appreciate, and it is growing …

AFTERWORD
... looking ahead ...

Space science and technology are moving forward ever more rapidly. Increasing understanding of our place in the Universe brings new insights into our origins and our existence, while the ability to leave the confines of planet Earth is opening up new and exciting opportunities for the human race. Less than a hundred years after the first liquid fuel rocket made its faltering leap into the air, we have taken great strides spacewards. People have floated free in orbit and walked on the Moon. There is nothing but time and money stopping us from venturing farther and staying away longer – perhaps even living on the Moon or Mars for extended periods.

Where this will lead in the next twenty, thirty or forty years – and who will go into space in that time – is anybody's guess. In the next few years at least, advances in space exploration are fairly predictable: although the pace of development in spacecraft technology is rapid, it still takes a few years of rigorous testing to move a new spacecraft from the planning stage to production and to its first flight. Building spacecraft is a time-consuming, skilled and risky business and, despite the initiatives of private financiers, there will be no cheap or easy access to space in the next few years. However, the commercialization of space is a growing force, pushing the technology forward and the costs down. Through huge investment and such initiatives as the X Prize, space will no doubt eventually become accessible to many more people over the next few decades. This will satisfy those frustrated with the slow pace of government-led space programmes that have purely scientific goals. The dreamed-of bases or civilizations on the Moon and Mars probably will be realized – perhaps in the lifetime of a reader of this book.

But for the foreseeable future, most of the world's population will be firmly rooted on Earth. At each step along the way to greater accessibility to space, it will be important for us as a species to ask why we want to go there, and who will benefit. Certainly the people who are investing in space futures will benefit – they are shrewd business people who will proceed only if they can be guaranteed a financial reward. Then there are the pioneers, who simply want to push human frontiers ever further forward. It is natural to want to do this, but we must always remain aware of who we are and why we want to go to space.

And it is worth remembering that the potential costs of space travel may be more than just the financial costs of building spacecraft. If space travel becomes routine, what will be the effects on the environment, both on Earth and in space? Will the dreams of the few leave the many behind? And most of all, are we really ready to push out into space when there are still so many problems to solve back here on Earth?

While the future of space travel is fairly predictable, the related sciences of astronomy, astrophysics and cosmology will no doubt continue to be full of surprises. For although theoretical predictions of increasing sophistication are being realized, we uncover more questions and fascinating phenomena that beg for answers. The modern instruments of astronomy can peer deeper into the Universe with increasing clarity, but there will always be more to learn and more to tantalize or to astonish us.

One thing is for certain: in the modern world, space science and space technology are co-dependent – they are locked in a permanent relationship. Space telescopes and space-station laboratories require funding, and they require efficient transportation. The intimate connection between private and public funding of ventures into space – where scientific missions ride on the back of commercial ones, for example – will probably become more common, as the Russian and American governments shy away from the huge investment they once poured into the space race, driven as it was by political goals.

There is one aspect of NASA's space programme that must not be forfeited in the new era of space exploration, and that is education – in the widest sense of the word. The spectacular images of the planets and their moons, of distant stars and galaxies, and of our fiery Sun, appear in books, magazines and newspapers and on television and the Internet. When made freely available to everyone in this way, they surely justify the large sums of money spent on space programmes. As well as being inspirational, space programmes can bring more down-to-earth benefits to everyone through the remarkable advances in science and technology. The costs of mounting space-based scientific operations such as the International Space Station are monumental, of course, and there are many causes that would benefit from such funding. However, there are many other areas of human activity that involve just as much money but that do not bring such benefits – the arms race and the worst excesses of consumer society immediately spring to mind. Another unique feature of modern space exploration is the level of international co-operation that is involved, and here, too, is a reason to continue to venture into space.

There will always be more mysteries to solve in the Universe – the journey towards understanding space will have no end. Along the way, we must not lose our ability to stand and stare at the majesty of it all. Galileo was aware of this when he wrote: 'The Sun, with all the planets revolving around it, and depending upon it, can still ripen a bunch of grapes as though it had nothing else in the Universe to do.'

TIMELINE

The development of space travel

1926 Rocket engineer Robert Goddard launches the first liquid fuel rocket.

1949 V-2-WAC-Corporal rocket becomes the first manufactured object launched into space.

1957 Sputnik I becomes the first manufactured object in orbit.

1958 NASA is formed.

1959 Luna 2 becomes the first probe to hit the Moon.

1961 Astronaut Yuri Gagarin is the first person in orbit.

1965 First spacewalks: Aleksey Leonov (Voshkod 2) and Edward White (Gemini IV).

1968 Apollo 8 is the first crewed spacecraft to orbit the Moon.

1969 Neil Armstrong is the first person to step on to the Moon (Apollo 11).

1973 Skylab is launched.

1976 Two Viking probes land on Mars.

1981 First launch of the Space Shuttle.

1986 Mir space station launched.

1989 Galileo space probe sets off for Jupiter and its moon.

1994 Clementine space probe discovers ice on the Moon.

1996 The X Prize, a competition for private reusable spacecraft, is announced. Also, VentureStar wins X-33 contract.

1997 Cassini probe sets off for Saturn, carrying 32.4 kilograms of plutonium.

1999 First part of the International Space Station in orbit.

2000 First crew board the International Space Station.

The development of astronomy and astrophysics

1543 Nicolaus Copernicus publishes his theory that the Sun is at the centre of the Universe.

1610 Galileo Galilei makes the first scientific observations of the night sky with a telescope.

1667 Isaac Newton publishes his Universal Law of Gravitation.

1838 Friedrich Bessel is the first to measure the distance to a star.

1861 Gustav Kirchoff and Robert Bunsen discover the composition of the Sun using spectroscopy.

1905 Albert Einstein publishes his Special Theory of Relativity.

1916 Albert Einstein publishes his General Theory of Relativity.

1916 Karl Schwarzschild works out black-hole theory using Einstein's general relativity.

1929 Edwin Hubble discovers the law dictating the expansion of the Universe.

1964 Arno Penzias and Robert Wilson discover the cosmic background radiation from the Big Bang.

1967 Jocelyn Bell discovers the first pulsar.

1971 Cygnus X-1 is proposed as a strong black hole candidate.

1974 Stephen Hawking suggests that black holes emit radiation.

1989 COBE (Cosmic Background Explorer) satellite is launched.

1990 Hubble Space Telescope is launched.

1992 V404-Cygni is discovered as most likely black-hole candidate.

1995 SOHO (Solar and Heliospheric Observatory) is launched into orbit around the Sun.

1997 Astronomers at the European Southern Observatory confirm the existence of a super-massive black hole at the centre of our galaxy.

1999 Hubble Space Telescope's Key Project Team concludes, and suggests that the value of Hubble's constant is 73 kilometres per second per megaparsec.

2000 First isolated black holes discovered.

SELECTED
BIBLIOGRAPHY

Apt, Jay, Michael Holfert, Justin Wilkinson and Roger Ressmeyer (ed.),
Orbit: NASA Astronauts Photograph the Earth, National Geographic
Society, Washington, 1996.

Begelman, Mitchell, and Martin Rees, *Gravity's Fatal Attraction: Black
Holes in the Universe*, Scientific American Library, Chicago, 1996.

Charles, Philip, and Mark Wagner, *Exploring the X-ray Universe*,
Cambridge University Press, Cambridge, 1995.

Couper, Heather, and Nigel Henbest, *Space Encyclopedia*, Dorling
Kindersley, London, 1999.

Ferguson, Kitty, *Prisons of Light*, Cambridge University Press,
Cambridge, 1996.

Henbest, Nigel, and Michael Marten, *The New Astronomy*, Cambridge
University Press, Cambridge, 1996.

Hufbauer, Karl, *Exploring the Sun: Solar Science Since Galileo* (New Series
in NASA History), Johns Hopkins University Press, New York, 1993.

Light, Michael, *Full Moon*, Jonathan Cape, London, 1999.

Padmanabhan, Thanu, *After the First Three Minutes*, Cambridge
University Press, Cambridge, 1998.

Thorne, Kip, *Black Holes and Time Warps: Einstein's Outrageous Legacy*,
W. W. Norton & Co., New York, 1994.

Turner, John, *Rocket and Spacecraft Propulsion*, John Wiley and Sons,
London, 1999.

Wilson, Robert, *Astronomy Through the Ages*, UCL Press, London, 1997.

INDEX

solar wind 79, 128, 129, 132, 151, 156, 157
solid-state phased array radar 105–6
Space Command 105, 115–16, 132
Space Environment Center 134, 154, 155
space junk 13–14, 91–121, 132–3
Space Shuttle 17, 18, 25–6, 30, 36–7, 100–1, 104, 112, 119
Space Surveillance Network 110
space-time 165, 166–70, 186, 187, 189, 218–19, 220–1
Space Tourist 46
space walks 92–4, 133
spectrograph 147
spectroheliograph 145, 152
spectroscopy 12, 142–3, 144–5
spiral galaxies 200–1, 211, 216
spiral nebulas 199–200
Sputnik program 27–8, 29
SR-71 38
starburst galaxy 186
Steady State theory 206
stellar black holes 182, 187
Stroupel, Eliyahu 134
Sun 8, 14, 123–58, 160, 161–2, 168–70
sunspots 136–41
Sunyaev-Zel'dovich effect 214, 216
super-massive black holes 182–6, 187
supernovas 11, 163, 173, 181, 210
system for nuclear auxiliary power 117

T

Tammann, Gustav 209, 210, 211, 212
Teledesic 107
telepresence systems 75–6
Telesat 130–1
Telestar 401 satellite 131
Thorne, Kip 171, 172–3, 179, 188, 193
Thunderbird 47
Tinsley, Brian 139–40
Tonry, John 212
tourism 13, 15–18, 40, 50–1, 63–90
Tsiolkovsky, Konstantin 22, 27
Tymon, Dave 106

U

Uhuru 176
ultraviolet 12, 147, 168

Ulysses 149
uncertainty principle 191
United Nations treaty 65
Uranus 203
Urie, Dave 34, 35

V

V-2-WAC-Corporal rocket 18–19, 24
V404 Cygni 159–60, 162, 171, 180, 182
Vanguard rocket 26, 28
Vaucouleurs, Gérard de 209
VentureStar 34–5, 36–9
Very Long Baseline Array 217
virtual particles 191
von Braun, Werner 23, 28
Voyager 45

W

Wallis, Michael 50–1
weather 14, 123–58
Wheeler, John 174
white dwarfs 161–2, 174, 212–13
White, Ed 91, 92–4
Whittaker, Red 75
Whitten Brown, Arthur 41
Williams, Lottie 114
Wilson, Robert 207
WIMPs 223
Wind 150, 156
wormholes 187–8

X

X-38 120–1
X-plane programme 31–40
X Prize 40–1, 42–7, 50, 67, 107, 236
X-rays 146, 147, 148, 149, 159, 168, 175, 176, 178, 180, 181–2, 186, 188–9, 224
X Van 46

Y

Yeager, Jeana 44–5
Yohkoh 150

Z

Zel'dovich, Yakov 174–5, 176, 183
Zwicky, Fritz 210, 223